North Molton Gold!
©Fred Harding

Foreword by Peter Stucley

(Great grandson of George Wentworth Warwick Bampfylde, 4th Baron Poltimore (1882-1965)

COPYRIGHT

First published in the United Kingdom

July 2012

Second Edition published December 2013

Copyright © Fred Harding

All rights reserved. No part of this publication may be reproduced, stored in a retrieval system or transmitted in any form or by any means without the prior permission in writing of the publisher, nor be circulated in writing of any publisher, nor be otherwise circulated in any form of binding or cover other than that in which it is published without a similar condition including this condition, being imposed on the subsequent purchaser.

ACKNOWLEDGEMENTS

This book would not have been possible without the help and support of the people listed below. I would like to thank:

Peter Stucley, the youngest son of Sir Hugh Stucley (6th Baronet), who has written the Foreword for this book. He also gave me permission to visit the Bampfylde mine upon whose land the site is located during my investigations.

Norman Grovier whose local knowledge of the mines and his tours of the Bampfylde mine have been most invaluable and inspirational. Also for the photos and various documents that he also provided for this book as well as his proofing the book.

Roger Burton for permission to use the letter containing the certificate of analysis by OMAC Laboratories Ltd that was sent to him by the surveyor Mr Hamilton who carried out samples of the Bampfylde mine on behalf of Feltrim Mining in 1989. Also, various photos and documents that he also provided for this book.

Phil Tonkins assistant at the Museum of South Molton.

The British Geological Survey for permission to reproduce under license the assessment, conclusions and recommendations from their report entitled **An Appraisal of the Gold Potential of Mine Dumps in the North Molton Area, North Devon** by D G Cameron, BSc, BA and D J Bland, BA. The survey was carried out in 1994. Reproduced by permission of the British Geological Survey © NERC. All rights reserved. CP12/053

PICTURE CREDITS

Unless stated otherwise, the photos and artwork in this book were created by the author Fred Harding and are copyright.

Third Party Photos and Drawings are acknowledged, permissions obtained and credits referenced.

Bampfield Mine Schematic based upon drawing supplied by Norman Govier. It has been edited and colourised by the author.

CONTENTS

Acknowledgements
Picture Credits
Foreword
BY PETER STUCLEY, GREAT GRANDSON OF GEORGE WENTWORTH WARICK BAMPFYLDE, 4TH BARON POLTIMORE (1882-1965)

Introduction

Chapter 1 - Fact or Fiction?
JOHN CALVERT (1853) | MORE DOCUMENTED EVIDENCE | SHIPPING EVIDENCE | A SPANNER IN THE WORKS | GOLD OR NO GOLD - THAT IS THE QUESTION

Chapter 2 - Gold at the Museum
ABOUT SOUTH MOLTON | REPORTS OF BAMPFYLDE GOLD | I VISIT THE MUSEUM

Chapter 3 - The Mines of North Molton
THE BAMPFYLDE MINE | BAMPFYLDE HISTORY | EXTENT AND LAYOUT OF THE MINE | THE DECLINE AND FALL OF THE MINE | REBORN AS THE PRINCE ALBERT MINE | GOLD FEVER | MANGANESE WORKINGS | LITTLE MINES WOOD | BRITANNIA MINE | CROWBARN MINE | FLORENCE MINE | NEW FLORENCE MINING COMPANY | NEW FLORENCE TRAMWAY | THE WAR YEARS

Chapter 4 - The Great Gold Fraud of 1853
THE BRITANNIA MINING COMPANY | THE DUPING OF JOHN MITCHEL | BENJAMIN MASSEY CERTIFICATION | LORD POLTIMORE SMELLS A RAT | MR FLEXMAN'S NUGGET | SIR CHARLES SHARPE KIRKPATRICK CONVICTED OF FRAUD | THE COST-BOOK PRINCIPLE | NO GOLD AT THE BRITANNIA MINE | THE PERKES MACHINE | THE SWINDLE DISCOVERED | THE POLTIMORE COPPER AND GOLD MINING COMPANY | JAMES COOK - AN EXPERIENCED PROFESSIONAL | THE OTHER DIRECTORS | CAPTAIN MOORSOM - CONSULTANT ENGINEER | POSITIVE GOLD REPORTS | DESPERATE MEASURES | THE BERDAN MACHINE TECHNOLOGY | RIPPED OFF? | THE END IS NIGH! | THE POLTIMORE COMPANY FOLDS | AFTERMATH

Chapter 5 - I Visit the Bampfylde Mine
LOCAL KNOWLEDGE MAKES ALL THE DIFFERENCE | AN INVITATION I COULD NOT REFUSE | THE ADVENTURE BEGINS | I MEET THE YOUNGEST SON OF SIR HUGH STUCLEY | ACCESS TO BAMPFYLDE MINE | THE CRUSHER HOUSE | NORMAN SHOWS ME SOME GOLD | THE ENGINE SHAFT | WHERE THE BERDAN MACHINES WERE LOCATED | END OF FIRST VISIT

Chapter 6 - Three Men and a Mine
NORMAN MEETS JOHN ROTTENBURY | KINDRED SPIRITS

Chapter 7 - The Bampfylde Mine Revisited
SHAFT No 3 | SHAFT No 4 | I SEARCH FOR GOLD

Chapter 8 - The Truth is Revealed

FELTRIM MINING SURVEY (1989) | BRITISH GEOLOGICAL SURVEY (1994) | BGS - ASSESSMENT | BGS - CONCLUSIONS AND RECOMMENDATIONS

Chapter 9 - The Icing on the Cake
WHAT IS THIS? | ANOTHER DISCOVERY | TENTATIVE GOLD TESTING | THE THIRD VISIT TO BAMPFYLDE MINE | THE MOMENT OF TRUTH

Chapter 10 - The Sting in the Tail
SOMETHING IS NOT QUITE RIGHT | THE PHONE CALL | FURTHER TESTS | THE ACID TEST | THE GLITTER TEST | THE HARDNESS TEST | I RETURN TO SOUTH MOLTON MUSEUM

Chapter 11 - There's Gold in Them Hills
WHERE IS THE GOLD TO BE FOUND? | THE ESTIMATED VALUE OF THE GOLD | A SMALL SCALE COMMERCIAL ENTEPRISE? | THE KNELSON BATCH CONCENTRATOR | THE ALL IMPORTANT QUESTION

Epilogue - 2013
THE PROPOSAL | THE SITE SURVEY | GOLD CONFIRMED | BOMBSHELL! | WHAT NEXT?

Bibliography and End Notes
BIBLIOGRAPHY | USEFUL WEBSITES | END NOTES |

About the Author

Other Books I have Written
NEPHILIM SKELETONS FOUND | BREAST CANCER: CAUSE - PREVENTION - CURE | EVOLUTION'S COUPE DE GRACE | THE TIMES OF THE GENTILES ARE FULFILLED

Career History

FOREWORD

Although North Molton is a village with a rich mining history, I have always thought it a shame that one finds so little information about it in the public domain. Since my involvement with the estate, I have encountered various clues as to what went on all those years ago, be it on the ground or speaking to some of the more senior parishioners in the village.

As the years go by, it has concerned me that we are less and less likely to have an accurate historical account of North Molton's mining history. I was therefore delighted to discover that someone has at last pulled all the facts together, at the same time dispelling some of the myths. Not only is Fred Harding's book a fascinating record for future generations, it is also a mineralogical thriller for today. I am sure that anybody with an association with North Molton will thoroughly enjoy reading "North Molton Gold!"

by Peter Stucley
Great grandson of George Wentworth Warwick Bampfylde, 4th Baron Poltimore (1882-1965) and the youngest son of Sir Hugh Stucley, 6th Baron of Affeton Castle, Devon.

INTRODUCTION

In October 2009, I sold my house in Aylesbury in Buckinghamshire and moved to Barnstaple in Devon with my family. Here I purchased a nice house with a large garden that overlooked woodland through which a babbling brook flowed. It was as if my new home was actually in the woods rather than on an estate where it really was situated.

Birds of all kinds, squirrels and huge dragon flies frequented my garden often and the woods are always filled with the songs of bird life everywhere. Throw out some bread on to the lawn and as if by magic seagulls will swoop from the sky forcing away the magpies, pigeons and black birds who had dared to venture to feed on the bread. I give the birds bird feed and fat balls instead which the seagulls do not like.

Close to the sea and a short walking distance to open countryside it would appear that I lived in an idyllic setting with a quality of life that promised to be all that I have ever dreamed, quiet and tranquil. However, all was not perfect. I had a problem, a big one. One cannot live in paradise without the means by which to support the kind of lifestyle that I had become accustomed over the years and for which I had worked so hard to maintain. Here in Devon I had no job!

It is a fact that jobs are few and far between in this part of the world, so unless I could get employment in the tourist industry or retail trade, probably part-time, my options were very limited. Besides those jobs did not appeal to me. To make matters worse, I was somewhat long in the tooth, age 58 at the time, and therefore unlikely to be employable. It did not matter how skilled and experienced I might be I was just too old.

I was very conscience of the job situation here in Devon prior to moving and I did have a solution. Throughout my career I had acquired extensive skills in web design and software development and I had just left my last employment as Software Development Manager, a position I had held for eleven years. I was very pleased when one of my software products called CMS (COSHH Management System) that I had developed for the company reached the finals of the Oscar like awards of the Information Technology industry - the British Computer Society Awards. This was in 2003.

It was quite natural for me to start my own business using the website skills I had learned over the years to encourage the local business community to have a website. This was my immediate short term aim but ideally I wanted to develop software because that would offer me better financial rewards in the future, and which once done, would only require limited support. This would then enable me to spend more time writing books such as the one you are presently reading.

Like everything in life Rome was not built in a day so starting from scratch it took a while to establish myself. However, my transparent pricing policy and the design of my websites were so good that word soon got around and by the end of 2011 I had completed over 40 websites.

In June 2011 I had the opportunity during a lull in website development to write some software. I had purchased a Kindle from Amazon. This an e-book reading device which has taken the publishing world by storm. As I said earlier it was my intention to write books and I have had two published so far through POD (Print on Demand) methods. This was still expensive so with the popularity of the Kindle and e-books in general, this I decided was the way to go.

For intents and purposes to publish a book for the Kindle it was just a question of writing your book, creating the cover and uploading it to KDP (Kindle Direct Publishing). Would that this was as easy as it sounds. I was amazed to discover that there was no software available that one could use to write Kindle books. Instead, one had to hope for the best with conversion programs and the results were often not very good. Seeing a niche opportunity and my personal need to be able to publish my own books for the Kindle platform, I developed a software program called Kindle Writer.

Kindle Writer enables authors who have limited IT skills to easily write books for Amazon's kindle devices. This book you are reading was created with the software and since its launch in August 2011 it has helped many authors all around the world to write their books. It is a very low cost solution and I have received many testimonials praising the software. However, this is not what this book is all about. It is just one link in a chain of events that would lead to the publication of this astonishing book.

It was against this background that events would conspire that would lead me on a journey of adventure and discovery, one that I had thought could not happen in this day and age, and it all began with a website that I had developed for international artist Ken Hildrew.

Ken Hildrew lives in New Molton, which is a village close to Barnstaple where I now live. He had been so impressed with the work that I had done for him that when the North Molton Parish Council and History Society mentioned to him that they wanted a website for the village, he recommended my services. So it was that in April 2012 I met with two representatives of the village and subsequently I tendered for the work. Not long afterwards I received a letter of acceptance and so I began building the website.

The website was to be larger than the ones that I usually developed because the entire village was involved in the project. I must say that it was a most interesting and pleasant experience to work with so many people and it was clear from the start that the village was a very close knit community, something quite rare to find these days.

In a little over two weeks the website was ready and it went live. I was gratified to learn that feedback from the villagers had been very positive and most complimentary. Just do a Google search for North Molton and the website appears on the first page.

The North Molton website colours and theme had been designed around the official New Molton Book that was published by the History Society, a copy of which had been given to me. It was then that the next link in the chain of events occurred that would set in motion the adventure of discovery that you are about to read. It was while developing the website using the book as my guide when a sentence in the book caught my eye. In a chapter discussing copper mining in the area I read: *"Small traces of gold are still found their today by specimen collectors"*.

Gold! What gold? With the television bombarding us daily with advertisements to sell our personal gold for high prices, here I was living in an area where gold could be found. Wow! I wanted to know more and this book is the result of my endeavours.

What you are about to read is an extraordinary story and I invite you to travel with me to the mines of North Molton where I tread the ground and explore the site where one hundred and fifty years ago it was a hive of activity, the centre of extensive mining operations that employed hundreds of people.

On our journey we will meet incredible people whose parents and grand-parents were involved in the mines and who tell extraordinary stories of what happened there. Conspiracies, intrigue, fraud and gold fever - it is all here and more. But what about the gold? Was gold ever found at the mines or were the tales of nuggets and reports of gold discoveries all part of an elaborate scheme designed to entice would-be investors to part with their money on false hopes of rich rewards. And if the stories of gold were true, what about today? Can we still find gold there now? Join me as we go in search of the fabled gold of North Molton and who knows what we shall discover together.

Chapter 1
FACT OR FICTION?

JOHN CALVERT (1853) | MORE DOCUMENTED EVIDENCE | SHIPPING EVIDENCE | A SPANNER IN THE WORKS | GOLD OR NO GOLD - THAT IS THE QUESTION

I must confess that I began my investigation into the gold of North Molton with a certain amount of scepticism. Why had I not heard about it before I asked myself? While it is true that I was a newcomer to Devon I certainly had not heard about it through any magazines or books that I had read in my extensive library which also included works on mineralogy. However, now that I lived in North Devon I reasoned that there should be a great deal of information readily available about the mines locally.

The obvious place to begin I thought would be to visit the museum of North Molton. There was just one problem, a big one in fact. The village did not have a museum. Today North Molton is just a quiet rural place that sits on the south western edge of Exmoor National Park, a short distance from the A361 link road. It has few amenities. A shop & post office, primary school, a couple of village halls, garage, sports club, pub and a bus service are all that caters for its thirteen hundred souls that live here. Upon visiting the village one would find it had to believe that mining had once operated from what appears to be, at first sight, a tranquil unspoilt part of the Devon countryside.

Already things did not bode well. Most places with an interesting past would have a museum but I guess being a village off the beaten track, nobody thought it worthwhile having one. However, there was a pub in the village called "The Miners Arms" one of the ten that once existed in the village that I was to learn later that it had catered for the miners refreshments in Victorian times. So at least this did confirm that there had been mining in the area.

With no museum at North Molton, I visited Barnstaple Museum instead, which was near to me. I reasoned that Barnstaple being the largest town in the area with North Molton only ten miles away that the museum was bound to have some information. I was wrong! When I spoke to one of the museum assistants she said they did not have anything on the mines of North Molton. Goodness, I thought. Here I was in the nearest town to North Molton, with a harbour that had access to the sea where probably ore from the mines were shipped and yet the museum had nothing on the mines. Incredible!

By now, and I had only begun my investigation and I was not getting very far. Then the museum assistant dropped me a lifeline. She said that the South Molton museum did have a display of some kind about the mines. That was good news because I was beginning to wonder if I was chasing after wind. I therefore resolved to visit the South Molton museum sometime in the near future. First, there was an easier route to obtaining information about the mining operations of North Molton - the Internet.

With access to the Internet, I next began to carry out my research through this medium. I soon discovered that there was quite a lot of information about the North Molton mines, but what was written was rather confusing. For example, there was the Bampfylde mine, Poltimore and South Poltimore mines, Prince Albert mine, Prince Regent mine, Britannia mine, Florence mine, New Florence mine and Crowbarn mine all apparently within the same general area. The confusion was cleared up later after doing some more digging, if you excuse my pun. It turned out that some of the mines had different names depending upon who owned them during a particular time of mining activity.

The following is a breakdown of the different names applied to the four main mines in the North Molton area. This should help you understand which mine I am talking about when I refer to them in the book.

> North Molton Mine = Prince Albert Mine = Bampfylde Mine = Poltimore Mine
> Crowbarn Mine = South Poltimore Mine
> Prince Regent = Britannia Mine
> Florence Mine = New Florence Mine

JOHN CALVERT (1853)

I continued with my investigations and visited the Internet Archive. This is a website that contains many old books out of print and after a search I did find found one book that reported gold at North Molton. This was **The Gold Rocks of Great Britain and Ireland** by John Calvert published in 1853.

THE
GOLD ROCKS
OF
GREAT BRITAIN AND IRELAND,
AND
A GENERAL OUTLINE OF THE GOLD REGIONS
OF THE WORLD,
WITH
A TREATISE ON THE GEOLOGY OF GOLD.
BY
JOHN CALVERT,
OF AUSTRALIA,
MINERAL SURVEYOR.

LONDON:
CHAPMAN AND HALL, 193, PICCADILLY.
1853.

Quoting from the book it said: *"The Exmoor, or Molton district, is remarkable as being that district in which gold has been moat wrought in Britain of late years, but, until now, little has been raised. The gold here was known in Polwhele's time, for he expressly states it was obtained from copper ores-f- at North Molton"*. Calvert further said, *"Some of the gossan ores of North Molton have been reported as yielding twenty-seven per cent, of gold. Mr. Massey found eleven ounces per ton in picked specimens from Poldmore."*

This was all very interesting. Richard Polwhele (6 January 1760 - 12 March 1838) was a Cornish clergyman, poet and topographer. This mean't that *"in his time"* gold had been reported - perhaps a hundred years before the publication John Calvert's book. It appeared that I was on the right track at last. Furthermore, in the book Calvert wrote:

> *"The Britannia Mine was first worked as a gold mine under the name of the Prince Regent Mine. It is near the river Mole, and lies north of the Poltimore Mine. The gold bearing lode is gossan, and it passes through into the Poltimore Mine, it is also supposed closely to resemble the latter in structure, and to run parallel with it about one mile to the north. It has yielded gold stones of great richness. The mine is worked for copper likewise. In 1853 the company contracted for the erection of works for*

crushing and amalgamating their auriferous gossan under the direcdon of Capt. W. Moorsom and Mr. Mitchell."

"The Poltimore Mine is south of the Britannia ; and between that mine and the village of Heasly Hill, near the river Mole, on the property of Lord Poltimore. The Poltimore was previously worked as the Prince Albert Mine, but no attempt was made to obtain gold. The mine contains a good copper lode. In the adit level, east, are old workings for two hundred fathoms in length, and it is conjectured that this lode was worked for gold, as there is there a gold gossan lode from twelve to fourteen feet wide ; and there are many hundred tons of gold bearing gossan dispersed over the surfce of the mine.

"Malachite is found here, as in our South Australian mines which have copper and gold lodes. The gold was discovered at the Poltimore Mine in 1852, by some gentleman who took away lumps of gossan to London for assay. This mine has now produced several pounds' weight of gold. The first piece was nearly three and a half pounds : namely, twenty-six and a half ounces from twenty tons of red gossan. Some of the red gossan yields one ounce seven pennyweights per ton, and the brown six pennywdghts per ton. The average yield of fifty-four tons eighteen hundred of dry ore was sixteen pennyweights per ton; the total being one hundred and two ounces five pennyweights. It is to be observed that this gold is nearly twenty-four carats and worth 83s. 6d. per ton."

The information in John Calvert's book was very useful, written as it was in 1853 a year after gold was found at the two mines mentioned - the Britannia and the Poltimore mines. However, this was only one report. I needed to find additional information from newspapers or journals in Victorian times that would 'add more meat to the bone' to substantiate the claims that gold mining took place at North Molton.

MORE DOCUMENTED EVIDENCE

I continued my search on the internet and before long I found a notice published in **The Gentleman's Magazine 1819** which was reproduced in **Blackwood's Magazine 1820.** which seemed to confirm what was being said about the gold being found at North Molton.

> **The Gentleman's Magazine 1819**
>
> —Native English gold has also been found lately in Devonshire, by Mr. Flexman, of South Molton. It occurs in the refuse of the Prince Regent mine, in the parish of North Molton; the mine was discovered in 1810, and worked for copper, but was discontinued in May, 1818. The refuse is a ferruginous fragmented quartz rock, and contains the gold in imbedded grains and plates. Gold has been reported to be found in some other mines in that neighbourhood.

> **Blackwood's magazine, Volume 6 1820**
>
> Native English gold has also been found lately in Devonshire, by Mr Flexman of South Molton. It occurs in the refuse of the Prince Regent mine, in the parish of North Molton; the mine was discovered in 1810, and worked for copper, but was discontinued in May 1818. The refuse is a ferruginous fragmented quartz rock, and contains the gold in imbedded grains and plates. Gold has been reported to be found in some other mines in that neighbourhood.

The notice simply said, *"Native English gold has been found lately in Devonshire. It occurs in the refuse of the Prince Regent mine, in the parish of North Molton; the mine was discovered in 1810, and worked for copper, but was discontinued in May 1818. The refuse is a ferruginous fragmented quartze rock, and contains the gold in embedded grains and plates. Gold has been reported to be found in other mines in the neighbourhood."*

Once again gold had been reported at the mines of North Molton, but considering the importance of gold I was surprised how little was said about it. One reads about the great gold rushes of California, Alaska and Australia during the nineteenth century, but here in Devon it seems there was barely a whimper. Perhaps, I thought, I should change tact and look at this from a different direction and perspective.

SHIPPING EVIDENCE

I reasoned that if gold had been mined at North Molton then where did it go? How was it transported and to where? There had to be some documentation such as a cargo manifest that described such transportation I reasoned.

It was quite a job to find, but in the end I did find a cargo manifest. Surprisingly it was found in a newspaper published on the other side of the world, in New Zealand, and in a 'Letter to the Editor' section. The magazine was the **Nelson Examiner and New Zealand Chronicle** of 1853 and quoting from the Plymouth Paper of the same year it described the following:

> *"GOLD IN DEVON" - The Albatross, 100 tons cutter, is to proceed from this port to Barnstaple with the first portion of the*

machinery connected with the Poltimore Gold Mine, near North Molton, which is being manufactured at the foundary of Messrs. Mare and Co., of Plymouth.

> GOLD IN DEVON.—The Albatross, 100 tons cutter, is to proceed from this port to Barnstaple, with the first portion of the machinery connected with the Poltimore Gold Mine, near North Molton, which is being manufactured at the foundry of Messrs. Mare and Co., of Plymouth. After landing the machinery, the vessel is to convey 100 tons of auriferous gossan from the Poltimore mine to Liverpool, where it is to be reduced in bulk at the works of Messrs. Rawlins and Watson, who have already tried a small sample, 'in which the existence of gold was perfectly clear, and these gentlemen desire to try a quantity before determining on the average yield. The question of "Gold in England" consequently now assumes a very important and interesting position, and the Poltimore Company act wisely in bringing the point to an issue, as regards profitable yield of the precious metal from gossan. Our talented townsman, Mr. Oxland, chemist and metallurgist, is engaged to test the gossan and, in addition, to examine the slag which is found near the Poltimore mine in some abundance; and so determine, if practicable, whether the ancients smelted these for copper or for gold, of which works there is not the slightest tradition, although the fact of reduction establishments having existed cannot for a moment be doubted from the slag which is met with in three or four places in the neighbourhood of North Molton, on the proerty of Lord Poitimore.—*Plymouth Paper*

Nelson Examiner and New Zealand Chronicle, Volume XII, Issue 592, 9 July 1853, Page 6

Here we read of the delivery of equipment manufactured in Plymouth that was to be shipped to Barnstaple where it was to be unloaded and transported to the Poltimore Gold Mine near North Molton. So far so good! The cargo manifest next said:

> *"After landing the machinery, the vessel is to convey 100 tons of auriferous gossan from the Poltimore mine to Liverpool, where it is to be reduced in bulk at the works of Messrs. Rawlins and Watson, who have already tried a small sample, 'in which the existence of gold was perfectly clear, and these gentlemen desire to try a quantity before determining on the average yield."*

From the manifest it was clear that the Albatross was tasked to take on-board 100 tons of gold-bearing rock from the Poltimore Gold mine and transport it to Liverpool for testing in order to determine the average gold yield. This all sounded very plausible and reassuring for my investigations.

There is more in the article that is worth recounting. It ends by saying how one of the Plymouth artisans was actively engaged in testing the gossam, I presume on the mine site:

> "Out talented townsman, Mr Oxland, chemist and metallurgist, is engaged to test the gossan and, in addition, to examine the slag which is found nerare the Poltimore mine in some abundance; and so determine, if practicable, whether the ancients smelted these for copper or gold, of which works there is not the slightest tradition, although the fact of reduction establishments have existed cannot for a moment be doubted from the slag which is met within three or four places in the neighbourhood of North Molton, on the property of Lord Poltimore."

I have to admit that by now I had become a convert in the belief that gold had been found at North Molton and by the sound of it in considerable quantities. Even so, I had a nagging feeling that something was not quite right. I could see that the year 1853 was a prominent year for gold mining at North Molton but then afterwards for some inexplicable reason, reports of the mining operations seemed to have faded quietly away. How strange! I was to soon find out why!

A SPANNER IN THE WORKS

So far I had been researching early documents related to the mining operations at North Molton and these indicated that gold was being mined there. It was when I began to look at reports from later years, that I came upon a startling revelation. It began with a simple sentence recorded in *The National Gazetteer of Great Britain and Ireland* of 1868. Talking about North Molton the paper wrote:

> "There are two copper mines which were formerly worked, and in 1840 a large nugget of gold was discovered, from which cause, in 1853, a sham gold mine was imposed upon the public."

A sham gold mine! What was going on? I then came upon another document shortly thereafter. This was *Micronews of the Canadian Micro Mineral Association* published in October 1984. Referring to *Notes and Gleanings* July 16th 1888, the writer of the article Garry Glen wrote concerning the Gold Mines of Devonshire, England the following after talking about a gold nugget in the possession of Mr. Flaxman of South Molton:

> *"However, this may be, the rumour of gold attracted the attention fo a body of speculators who, about 1850 obtained leave from Lord Poltimore, and formed a company to work for gold. One Charles Heneage appears to have taken the lead in the formation of the company what he called 'The Poltimore Gold Mine'. No doubt it was just at this time that there appeared the nugget now in the Albert Memorial Museum, which bears the inscription 'found in a trappean rock at Poltimore.' Without going so far as to say this nugget was actually put into the mine at North Molton, suffice to remark that such things have been done, and that fine specimens like that in the Museum are not usually found in trap deposits. However, the nugget served its purpose, and to use the incisive words of the Directory [Devonshire Directory 1986], a "great fraud was got up in 1853, and a sham gold mine foisted on the Public"."*

I was stunned by the revelation of the fraud that took place in 1853 but when I had recovered from the shock and applied logic to the matter I became aware of an important fact. Long before the fraud had been perpetrated, historical records had reported the finding of gold at North Molton especially at the Bampflyde mine. That being the case I could not accept that these earlier reports could be fraudulent too. By now from what I had uncovered, I was convinced that there was gold to be found at the Bampfylde mine at least.

GOLD OR NO GOLD - THAT IS THE QUESTION

I was now stuck! What should I do now? I needed tangible proof, something to confirm that gold had been found at the Bampfylde mine. But where should I look? The local library seemed the obvious place to look but then I remembered that I had not as yet visited the South Molton Museum where it was said that there was an exhibit about the mine. Could there be any evidence there I wondered.

I surfed the internet and took a look at the South Molton Museum website. The museum had a mineral section and when I read what was said about it my heart missed a beat with excitement. A short paragraph read: *"The bulk of our minerals displayed are from the Rottenbury*

Collection from mines in the area mainly consisting of copper, iron, silver, lead, and gold specimens." I did not know who Rottenbury was at this time but here was a definite reference to gold specimens that had come from mines that were local. Could this mean the mines of North Molton? There was only one way to find out.

Chapter 2
GOLD AT THE MUSEUM

REPORTS OF BAMPFYLDE GOLD | I VISIT SOUTH MOLTON MUSEUM

South Molton is a municipal borough, and market town, situated on the bold western band of the river Mole, which falls into the Taw about six miles to the south. It lies about three miles south of North Molton and about 11 miles E.S.E. of Barnstaple. The town centred around the church and in the Middle Ages a square was laid out as a new market place.

South Molton has been a thriving town since the earliest days of recorded history, Gilbert de Turberville created the borough about the year 1150. The town's early wealth came from its importance as a centre for the wool trade, but this declined in the eighteenth and nineteenth centuries and South Molton's role changed to a transport, administration and service centre. It flourished as a market town and expanded with the building of the Town Hall and Assembly rooms and the introduction of schools and many more local businesses.

Tourism came with the motorcar and the trains, the latter unfortunately axed in 1966 by the infamous Dr Beeching when the line between Barnstaple and Taunton was closed. Even so, tourism continues to flourish and plays a very important part in the economy of the town and the surrounding area. The town is situated in an ideal central position within easy reach of Exmoor, Dartmoor, the rugged north coast and the beautiful sandy beaches to the west. The town has a population of about 5000; with its square still surrounded by small local shops unspoilt and much the same as it has been for years. Market day is Thursdays with an extra pannier market on Saturdays.

The South Molton Museum was founded in 1951. It was initially housed at No 1 East Street in premises provided by the Borough Council. The collections were built around a core of artifacts accumulated by Mr. W. Webber, the then town bailiff, and they were later moved to their current housing in 1966. The museum is now located on the ground floor of the town's 18th century Guildhall, which was completed in 1743 and is situated in the town square.

Guild Hall, South Molton. The museum is on the ground floor.

REPORTS OF BAMPFYLDE GOLD

With the revelation that there were samples of gold from local mines in the museum, I contacted Jenny Yendall, the Museum Access and Collections Officer at the Museum. I asked Jenny if the gold samples had been found at the Blampflyde mine and she said that they had been as well as from other mines in the area. However, she could not be more specific as to which came from what. She said that if I wanted to visit the museum they did have some information on the mines and the minerals that were found in each of them.

Jenny further said that her colleague, museum assistant, Phil Tonkins, was a local historian and that he could fill me in with the details about the mines and the Rottenbury collection. She said that he would be happy to meet with me if I so wished. Needless to say I jumped at the chance. So arrangements were made and on the 31st of April 2012 I drove down to South Molton to meet with Phil in hopeful anticipation of what he might reveal.

I VISIT THE MUSEUM

On my arrival at the museum I was immediately made welcome, and Phil showed me the display of the minerals that had interested me. It had been donated to the museum by John Rottenbury, a local farmer who had become a well respected qualified geologist but who had died not a few years ago. I had observed that John's unpublished work *Geology, Mineralogy and Mining History of the Metalliferous Mining Area of Exmoor* that enabled him to obtain his Ph.D. from the University of Leeds in 1974 was often quoted by historians who have been researching mining in the Exmoor area.

Phil Tonkins showing me the Rottenbury Collection at the museum

The display was locked up, but Phil was able to procure the keys and before long I had in my hands a rock containing actual gold found at the Bampflyde mine. I was elated. It was true! Gold had been found at the mine and here was the proof that I had been seeking.

The rock that I held was quite large, about the size of the palm of my hand. On the edge of the rock was a thin layer of gold that glinted in the light from a small window above the display cabinet. I asked Phil if it was real gold or fool's gold and he confirmed that I had the genuine article in my hands. Tests had confirmed that the rock contained gold.

Rock from the Rottenbury Collection showing gold on the surface

There were two other smaller rock samples in the cabinet, which also were said to contain gold on their surfaces. For me this was further evidence which supported the written records that I had thus far obtained that said that gold had been unearthed in the North Molton parish.

The other two rocks from the Roddenbury Collection showing gold on the surface

Phil had been most helpful but his knowledge was limited to what was displayed. It was clear that I needed to find out more about the mining operations that took place at the Bampfylde mine. More research needed to be carried out. In the meantime, I thanked Phil for his kindness and departed, happy in the knowledge that I had seen tangible evidence that gold had been found at North Molton.

On parting Phil mentioned that one could not visit the mine unless one had permission from the owners of the site namely the Stucley family. He suggested that I should contact Peter Stucley who was the youngest son of the family and who owns the land upon which the Bampfylde mine was situated. This I planned to do in a month or two when the weather had improved but little did I know that I would meet the Peter much sooner than I thought and under very favourable circumstances.

Chapter 3
THE MINES OF NORTH MOLTON

THE BAMPFYLDE MINE | BAMPFYLDE HISTORY | EXTENT AND LAYOUT OF THE MINE | THE DECLINE AND FALL OF THE MINE | REBORN AS THE PRINCE ALBERT MINE | GOLD FEVER | MANGANESE WORKINGS | LITTLE MINES WOOD | BRITANNIA MINE | CROWBARN MINE | FLORENCE MINE | NEW FLORENCE MINING COMPANY | NEW FLORENCE TRAMWAY | THE WAR YEARS

The four mines of North Molton - Bampfylde, Crowbarn, Britannia and Florence.

There are a number of mines in the parish of North Molton surrounding the little hamlet of Heasley Mill. These are the Bampfylde mine later known as the Prince Albert and then renamed Poltimore mine, the Crowbarn mine also known as South Poltimore, Prince Regent mine later known as Britannia mine and Florence mine later known as New Florence. Of these the Bampfylde mine has been the most important one that has been worked so I will begin by describing this one first.

THE BAMPFYLDE MINE

As one travels northward from Heasley Mill along a narrow country road and before one reaches a small bridge called Mines Bridge, you would have passed through the Bampfylde mining area that stretches on both sides of the road. The river Mole runs through the site, its waters was used to drive the machinery of the mine, water power was used extensively. Today, there is not much to see from the road as the trees and scrub of woodland obscure observation and besides the area is fenced off. It is private land owned by the Stucley family and trespassing is forbidden.

It is when the site is seen from the air that one comes to appreciate the massive scale of the mine workings. Great scars on the land caused by the mining operations are clearly visible on both sides of the river Mole.

BAMPFYLDE MINE (also known as North Molton, Prince Albert and Poltimore)

BAMPFYLDE MINE HISTORY

Bampfylde mine was for most of its history a copper mine but manganese and iron was also worked within the mining area a little to the south. Later, and the subject of this book, gold was mined or so it was said. The mine has been closed now for a very long time, over a century and very little traces of the mine workings or buildings now exist. Trees and foliage now cover much of the land that had once been scarred by the mine workings, but nature has not as yet been able to cover over the huge heaps of pinkish red gossan that litter the area. Nothing can grow on these mounds of sterile rock.

There has been a long history of mining in the area and it is said that mining here goes way back to Roman times. Later the Bampfylde mine was known by the name of 'The King's Copper Mine' during the reign of King John but the earliest documentary reference to the mine known is recorded in 1346. It was from this date onwards throughout Medieval to Victorian times that the Bampfylde mine (known then as the North Molton mine) came into prominence with copper being extensively mined here.

By 1699 the mine was supplying 50% of the input to the UK's smelting industry. This is born out by the *Bulletin of the Peak District Mines Historical Society, Volume 11, Winter 1991*, where it is reported that in the first half of 1695 106 tons of ore from the mine was shipped out of Padstow and in the following year this had increased to 424 tons. In 1699 over 619 tons was shipped out of Barnstaple and Bideford. However, by 1701 shipments began to decline dropping to about 322 tons and a year later this had dropped to only 150 tons.

Copper mining had been very profitable at Bampfylde while copper deposits lasted. Hence, we read in the *The History of Devonshire, Book III - Outlines of the Geology, Physical Geography and Natural History of Devonshire* by Rev. Thomas Moore that:

"The ore of North Molton mine was plentiful in 1729, and sold then at £6.10s per ton, a good price at the time."

Copper mining continued at Bampfylde mine for about another fifty years and it still played an important role in providing copper for the UK industries. In *Exmoor's Industrial Archaeology* by Michael Atkinson and published in 1997, the author said:

"Between 1696 and 1698 over 4,500 tons of high grade (30% -

40%) copper ore was raised, with further reports stating that between 1724 and 1773 roughly 45 tons of ore was sold every month."

As the nineteenth century approached mining at Bampfylde had long past its peak of productivity and profitability. By now the mine had become a huge sprawling complex with four main shafts being actively worked. Three of the shafts were located high up on the escarpment west of the river Mole. These were Bampfylde Shafts No 3 and No 4. Two smaller shafts - Hands Shaft and Footway Shaft - were also to be found in the same area further west.

East of the river was found the Main Shaft also known as Engine Shaft. This was the oldest shaft of the facility and that is why most of the mining operations, buildings and machinery was located here. Lying in the river valley close to the Mole river it could easily be supplied with an endless supply of water power. And further east still Field Shaft was dug.

EXTENT AND LAYOUT OF THE MINE

It is hard to imagine the scale of the mining operations at the Bampfylde until one takes a look at the geographical area of the site and its cross section. It was huge!

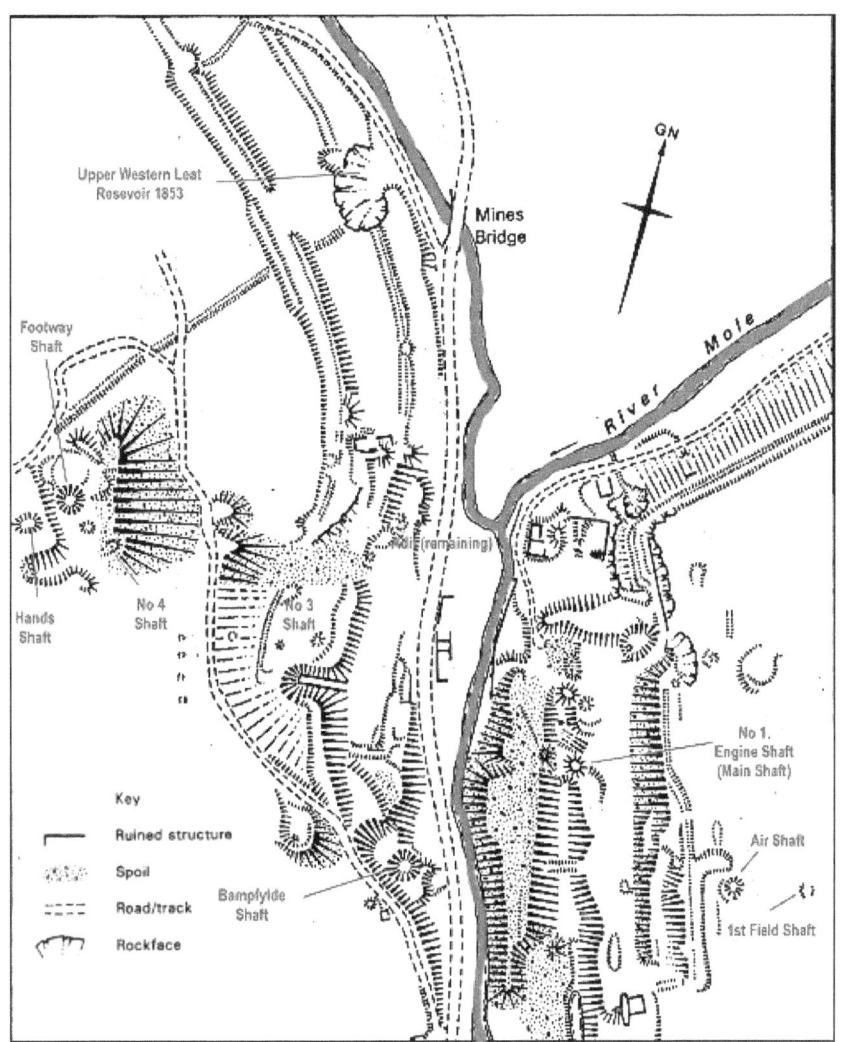

The mine shafts are described in the 1856 Cornish and Devonshire Mines found in the *Post Office Directory of Cornwall*.

The Bampfylde lode runs east and west, and dips south about 2 feet in the fathom; it consists of quartz and grey sulphate of copper. The known crosscourses are four, clayslate. There have been five shafts sunk. The No. 1 shaft is sunk 60 fathoms. There are the following levels in this shaft:-The 20-fathom level is driven 120 fathoms; the 30-fathom level is driven 130 fathoms; the 40-fathom level is driven 70 fathoms; the 60-fathom level is driven 10 fathoms south. The No. 2 shaft is sunk 60 fathoms. There is the following level in this shaft:-The 20-fathom level is driven 20 fathoms. The No. 3 shaft is sunk 70 fathoms. There are the following levels in this shaft: The 20-fathom level is driven 30 fathoms; the 30-fathom level is driven 50 fathoms.

The No. 4 shaft is sunk 20 fathoms. There is the following level in this shaft: -The 20-fathom level is driven 20 fathoms. The No. 5 shaft is sunk 10 fathoms.

All those statistics may be difficult to visualise but fortunately Norman Govier of whom I will speak more about later had in his possession a schematic of the mine that was published in 1868. This he gave to me for the book and as one can be see from the diagram the shafts had been greatly extended by this time. The diagram gives a very good idea how extensive the mining operations had been before they were finally closed not long afterwards.

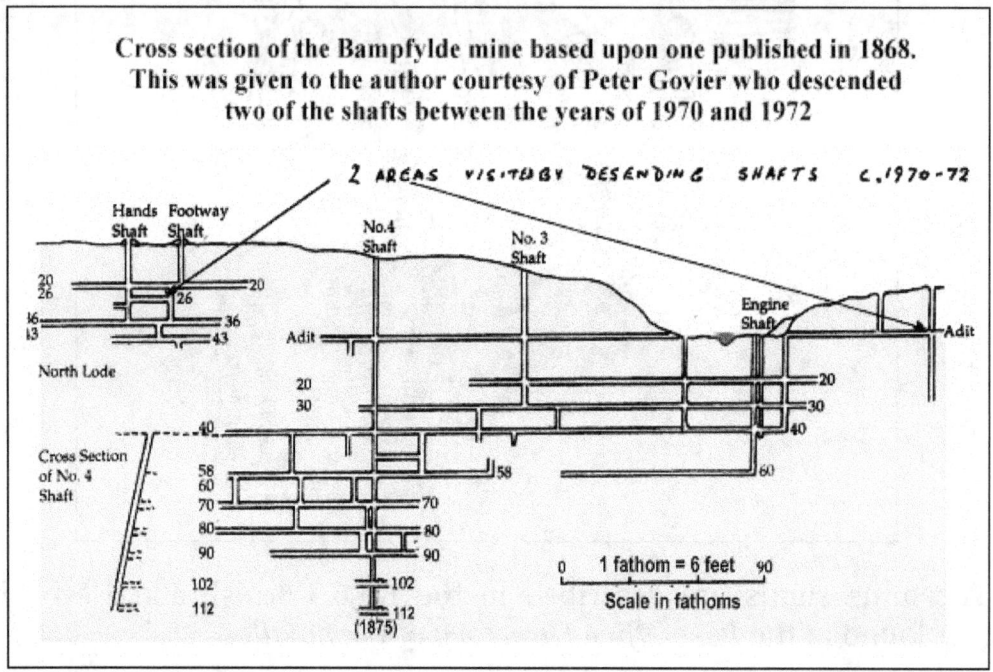

One picture, a photograph, has survived the ravages of time to show what the mine actually looked like when it was being worked. The picture is on display in the South Molton Museum but it is in poor condition, faded and brown with age. It is believed to have been taken around 1870. With permission from the museum I photographed the picture and upon returning home I digitally enhanced and colourised it for this book to bring it back to life.

A picture of the Bampfylde mine that I have digitally enhanced based upon a photo taken in 1870. The original photograph is on display at the South Molton Museum

In the picture you can see the engine shaft being worked and in the distance to the left is the "Captain's house". From here the mining operations were overseen by the mining foreman called "The Captain". This is a strange term but as William Tuck, c.1880 says in his *Reminiscences of Camborne*.

> "To a stranger coming to Cornwall, and visiting the mining district, he would be much surprised by hearing so many addressed as Captain, for in other parts of England the Army and Navy are only so honoured, but in this county anyone who occupies the post of overseer in tin or copper mining is dubbed captain.

In the picture one can also see smoke coming form the Captain's House chimney. Along the right hand side of the picture there is a rail track where wagons of ore were once moved by ponies ready to be transported to North Molton village, the distribution depot for shipment to the ports of Barnstaple or Bideford.

THE DECLINE AND FALL OF THE MINE

It was only a matter of time before the amount of copper ore dug from the Bampflyde mine dwindled to a point when it became uneconomical to process. Deeper shafts needed to be constructed and some of the shafts were already over a hundred fathoms (six-hundred feet) deep and were constantly flooded with water that needed to be pumped away. In addition, labour costs were increasing as were the transport costs for shipment through Barnstaple and Bideford.

To make matters worse there had been some accidents at the mine as extracting the copper was becoming increasingly hazardous. And of course, there was the license fee to Baron Poltimore that still had to be paid. The mine could no longer be financially sustained and in 1778 it officially closed. However, some mining did take place sporadically. Hence we read for example in *White's Devonshire Directory of 1850* when referring to the Bampfylde mine:

> "An old copper mine which had been closed many years was re-opened in 1813, but was soon abandoned."

REBORN AS THE PRINCE ALBERT MINE

In 1840 a group of entrepreneurs, one being Robert Backwell of Devonport near Plymouth, obtained a lease to mine copper from Lord Poltimore at one fifteenth royalty. The company that was formed was the **Prince Albert Company** and hence the Bampfylde mine was reopened as the Prince Albert mine. Unfortunately, there was not enough copper easily got at to make the business a profitable concern and so the company ceased operations five years later in 1845. However, it was also during this period that some gold was said to have been found in the slag heaps but there is no record of actual gold production by the company taking place.

Although the Prince Albert Company had ceased trading Robert Blackwell thought he saw an opportunity that might make him rich. With gold being found at the mine he approached the Treasury for a licence to work minerals covered by Crown privilege. He was to be disappointed.

While it was true the Mines Royal Acts of 1689 and 1693 had removed Crown privilege from silver and gold combined with lead, copper or other ores Robert Blackwell was disappointed to discover that gold in its free state, as found at the Bampfylde mine was still Crown property. Faced with considerable costs if successful with his bid he did not pursue his gold mining idea any further. He was not to know that 150 years later others would seize the same opportunity that he had declined to pursue. World events would one day conspire to create a flurry of activity at the mine and give it a new lease of life.

GOLD FEVER

In 1849 the world was gripped by "Gold Fever" as people from all over the United States and the world rushed to California in the hope of striking it rich. The California Gold Rush as it became to be called brought fame and fortune to many ordinary people, and at first gold was easy to find. All one needed it seemed was a knife, pick, shovel and a pan. It was said that gold nuggets could be easily pried from rocks and dirt shovelled from creeks and rivers could be swirled in a pan to find gold in significant quantities.

National newspapers around the world had a field day describing the rags to riches stories of those who had got lucky and many people were persuaded by the stories to leave their homes and families and try to seek their fortunes on the gold fields of California. On arrival in the chaos of the hub and bustle of the new mining towns springing up all over the place the newcomers were to find that things were not as rosy as they had been led to believe.

While it was true that in the early days gold could be found fairly easily, by 1853 this was not the case. The hundreds of prospectors that had managed to get in on the act at the beginning had made a killing, but land claims by them meant that newcomers had to make do with land that was less likely to contain much gold if any at all. The newcomers had arrived thinking that gold was just lying around on the ground just waiting to be picked up. They were faced with instead a hard reality check that this was not true. They were just not ready and ill equipped for the hard work necessary to mine for gold. However, all was not lost. There were other riches to be had ready for the taking and they grabbed it with as much enthusiasm as they originally had when they came to this foreign land.

It turned out that most prospectors were previously storekeepers, cooks, carpenters, teachers, farmers or had some other trade before heading to California to seek their fortunes on the gold fields. Hence, many of them made their fortunes by selling supplies and services to miners instead. This brought economic prosperity to California. Farms, ranches, stores, restaurants and other businesses that grew to serve the miners continued to take advantage of California's rich agriculture and thriving industry and commerce.

While all this was going on here in England some speculators in London saw an opportunity to jump on the Californian Gold Rush 'band wagon' and cash in on the stories of rags to riches described in the national press almost daily. But they had a problem. Nobody was interested in the so called reports of gold found in the mines of North Molton. Such reports were viewed with considerable scepticism, fairy stories with little evidence to back them up.

"Gold was first discovered at the North Molton Copper Mine (better known by its later title the Bampfylde Mine) sometime before 1785, but initially there is no evidence that it was anything more than of academic interest," says Peter Claughton, Hon. University Fellow - College of Humanities at the University of Exeter, Department of History.

Undaunted be the lack of interest and knowing that there was a long history of gold finds at the North Molton mines, the speculators conceived a daring plan. They needed to up the stakes and make the idea of gold mining at North Molton as an exciting prospect just as the Californian Gold Rush had been. So they arranged for a gold nugget to be planted at one of the mines, the Prince Regent mine. It had the desired effect as we shall learn in the next chapter.

MANGANESE WORKINGS

Before closing on the subject of the Bampfylde mine it should be mentioned that although copper and then later gold were the primary motivation for mining at Bampfylde, there were other minerals that were mined here too. For example, on the southern boundary of the mine manganese and iron was mined. Unfortunately, there is virtually no information concerning this mining operation and all that remains today to testify of its existence are the ruins of a small Crusher House with a waterwheel pit alongside.

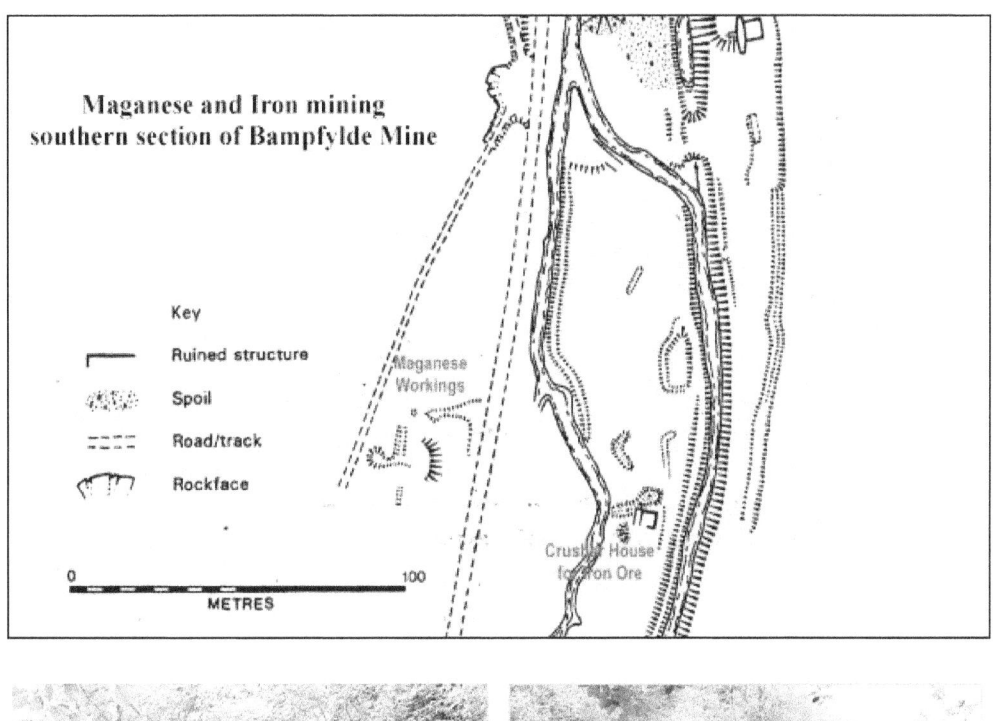

Manganese workings: Left: Ruins of Wheel and Crusher House. Right: Spoil from the Manganese mine

LITTLE MINES WOOD

A little north of the main Bampfylde mining complex is Little Mines Wood. Here some construction in connection with the Bampfylde mine was carried out. This was called **Upper Western Leat** and it was constructed in 1853.

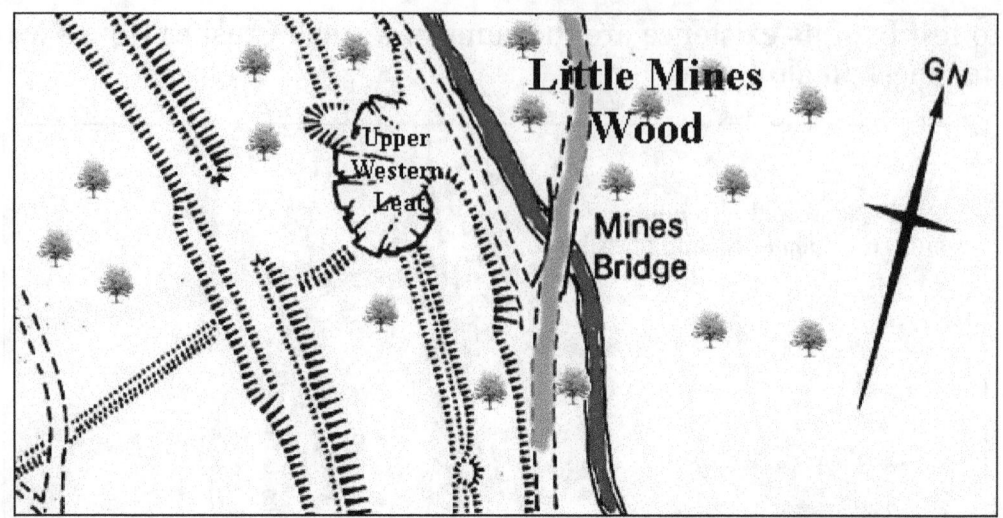

A leat (also lete or leet, or millstream) is the name, common in this part of England and in Wales, for an artificial watercourse or aqueduct dug into the ground, especially one supplying water to a watermill or its mill pond. Such was the case here. The Upper Western Leat was built to feed the 50' diameter 6½' breast pumping wheel that was being built on the Western bank of the river Mole for the purpose of powering the pumps in the engine shaft.

We can read about this wheel and the leat in a letter to the *Mining Journal* in February 1855.

> "....... There is a new wheel, not yet completed .... fixed on the face of the hill on the west side of the mine and about 40 fathoms distant from the pumping shaft .... to be connected by a horizontal beam working on supports at about 40 feet above the level of the road. .... the water to drive this new wheel is to be brought from a distance of more than 1¼ mile by a new leat which has been cut for the purpose on the face of the hill ..."

Where the 50' diameter 6½' breast pumping wheel was built

The wheel was never completed. It was built in the wrong place and was poorly constructed. All that remains today to show that it ever existed is an overgrown vertical slot dug into the slope going up towards where No 4 shaft is located on the western bank of the river.

As for the Upper Western Leat it was enlarged in the 1860s to form a reservoir which was used to providing a continuous supply to the water powered winder for No. 4 Shaft before the mine was finally closed.

BRITANNIA MINE

We now come to the second mine in the Heasley Mill area. A little north-east of the Bampfylde mine and about a mile away is the location Higher Mines Wood. It is here that the infamous Britannia mine was worked.

CAPTAIN JOSEPH ODGERS

It was about 1806 when we first hear about mining at Higher Mines Wood. It was begun under the agency of Captain Joseph Odgers. He drove an adit about 40 fathoms and sank a shaft for about 4 fathoms (twenty feet), but this became flooded and he did not have the means pump the water out. However, he had managed to excavate some copper ore that was sold between £7, 10s to £8 per ton so he was able to convince a group of business men to buy the mine for £600. We can read this in a recent discovery of a large quantity of lecture notes by William Thomas written in a neat hand on foolscap sheets and headed *Lecture Notes on Mining 1892-8*.

The Introductory lecture in the notes for September 1891 begins thus:

"Capt. Joe Odgers... on one occasion sold a sett to some parties for £600 & they came down to go over the ground. At last one of the purchasers said 'Well Capt. Joe this is all very cheering & now show us the lode'. 'Lode!' he replied 'Lode! my dear man, the sett would be worth ten times as much if there was a lode in 'un'."

The mine to which Captain Odgers referred was none other than the Britannia mine. In the prospectus by the Britannia Gold and Copper Mining Company written in 1952, we read the following:

"The BRITANNIA Mine is the property of Lord Poltimore, and is situated about seven miles north of South Molton, towards Exmoor, on the banks of the Mole. It was first worked about forty-years since, for Copper, under the agency fo Capt. Joseph Odgers, but with very inefficient meands, and after driving an adit about 40 fathoms and sinking a shaft 3½ or 4 fathoms they were driven out by water, having no power to fork it......Subsequently, the work was worked by an association of London and Plymouth gentlemen...."

It would appear that it was to these 'London and Plymouth gentlemen' that Odgers had sold his set for £600 as aforementioned.

AS THE PRINCE REGENT MINE

It would appear that the people who had brought the rights to mine copper at the Britannia mine were the same people who worked the mine under the name of Prince Regent. According to the prospectus aforesaid we read:

"...but chiefly owing to the failure of a then eminent London banker, who was one of the principle Shareholders, operations again ceased. It then passed into the hands of two gentlemen, who worked it individually on a limited scale and it is from these gentlement that the present proprietors have purchased the remainder of the lease, which has about eight years to run"

By the time **The Britannia Gold and Copper Mining Company** came on the scene and reopened the mine in their name in 1852, the mine had already had a somewhat chequered and dubious history for copper production. It certainly could not compare to the Bampfylde mine further south either in the equipment used or the copper extracted.

The new company inherited workings that comprised no more than one adit (an entrance to an underground mine which is horizontal or nearly horizontal, by which the mine can be entered, drained of water, and ventilated). The adit was 81 fathoms (426 feet) long and there were only two shafts, one of which was 10 fathoms (60 feet) deep drained by a 25 foot water-wheel. However, this did not matter to the people behind the **The Britannia Gold and Copper Mining Company**. They had other plans for the mine and mining for copper was not one of them. It was to be an instrument of a scam and you can read more about that in the next chapter.

CROWBARN MINE

Travelling south from Heasley mill and away from the Bampfylde mine we soon reach Crowbarn Wood which straddles either side of the road. It is here that Crowbarn mine can be found. It was also known by the name of South Poltimore mine.

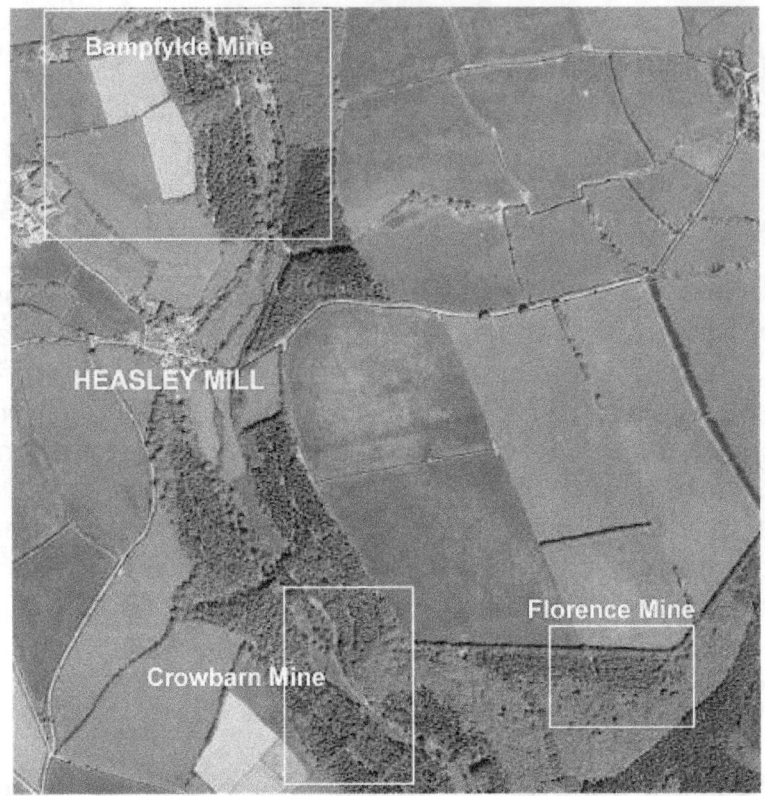

There is very little information still remaining concerning this mine. There is evidence of iron mining having take place here, but this is probably related to activities of the Florence mine to the west. Later, horses for the tramway serving the Florence mine was stabled here.

When gold was being searched at the Britannia and Bampfylde mines to the north it made sense to look in the southern section of the mine sett too. However, as was found in a recent survey by the British Geological Survey of 1994 no gold was found here. When the Gold swindle was discovered at the Britannia mine in 1853 resulting in the closure of that mine, followed shortly by the Bampfylde (Politmore) mine, the South Poltimore mine closed too. This was about 1859.

We next hear about the mine when it was opened again in 1873 when it was reworked for extracting manganese and iron as their were small but rich veins discovered here. However, mining only went on for about ten years and the mine was finally closed in 1884. It has remained closed ever since. Today, there is very little evidence to show that mining activities ever took place here.

FLORENCE MINE

South-East of the Bampfylde mine can be found the **Florence Mine**.

This mine lies within Radworthy Cleave, a remote, wooded and steep-sided combe on the southern edge of the moor.

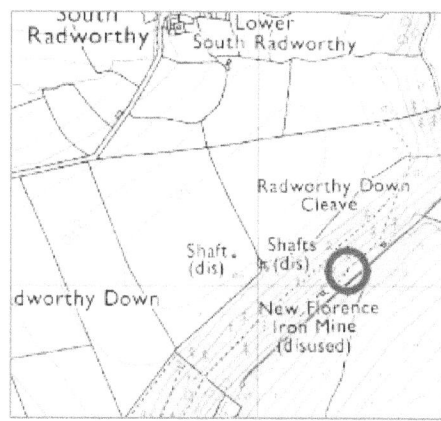

Location of (New) Florence Mine

Very little is known about the mining operations at the Florence mine. According the National Archives and records held at the North Devon Record Office, in 1868 George Bush of Wiveliscombe, Somerset leased the land from James Harris of Bittadon for 21 years in order to mine iron ore and manganese. He formed a company called the **West of England Iron Ore Company Ltd** with a surveyor called John Leversedge of Taunton. In the same year they were licensed to mine for iron ore or iron stone at Grit Down and Holdstone Down.

George Bush then persuaded a group of London entrepreneurs to join him in his enterprise to mine iron at the mine. These were two brothers Sydney Hawkins and William Bailey Hawkins who were iron agents, and two civil engineers Edward Woods and Julian Horn Folme. An investor Christopher Robins also joined the management group. As a result in 1871 the **Florence Mining Company** was formed and mining operations began in July of that year.

The workings were on two principal lodes. North and South, running approximately west-south-west to east-north-east. There are, in addition, a number of outlying adits and pits which appear to be more than exploratory workings. The South Load was the richest and it is here where most of the mine workings concentrated. A shaft was sunk 46 feet below the valley bottom and a cross cut driven to the load.

The enterprise was an immediate success and from 1873 to 1875 records show that 38,386 tons of iron ore had been extracted making it the most productive iron mine in the Exmoor foothills.

THE NEW FLORENCE MINING COMPANY

By 1879 output of iron from the mine had dropped considerably. The **Florence Mining Company** by then had got into financial difficulties due to a reduction in the price of iron ore and with the expense of mining operations that was by now giving poor returns. With still another eight years remaining on the lease the company went into liquidation but another consortium of business men took up the remainder of the lease and the company was renamed the **New Florence Mining Company** for a nominal capital of £1000 and the mine became known as the New Florence Mine.

NEW FLORENCE TRAMWAY

A narrow guage railway line was built in 1874 called **The Devon and Somerset Railway** and it connected to Taunton and Barnstaple. A branch of the line by-passed North Molton and at Brinsworthy it divided. One line went to South Molton while the western branch formed a tramway that was connected to the New Florence mine. Its overall length was some 5.5kms. The line was never completed to Heasley Mill and the mines used pack horses and carts to carry the ore to Crowbarn and the tramway.

Part of the tramway survives at its junction with the Crowbarn Mine Tramway, where it is a sharply defined, flat-topped embankment 0.9m high. North of Brinsworthy Bridge it has been disturbed and eroded in places by the erratic course of the unnamed stream. At several places it crosses and re-crosses this stream, and here attempts have been made to confine the stream with short sections of roughly coursed walling. In addition, lengths of tram rails still in place have been used as the basis for a makeshift bridge to carry the tramway over the stream but most of the course of the tramway is now impossible to follow in places.

There are still the remains of a loading bay fronting the tramway and adjacent to other structures associated with the New Florence Iron Mine. The main line continued for some 200m north-east of the centre of the complex, and ended alongside a massive spoil heap issuing from one on the principal adits.

Many of the miners were said to take advantage of the tramway to travel to North and South Molton to spend their wages in the public houses and get back to the mines in time for the next shift.

THE WAR YEARS

When the lease ran out in 1888 the **New Florence Mining Company** went into liquidation. There was some minor mining after this but everything came to an end in June 1894.

Several piecemeal attempts were made to reopen the workings in the 20th century. **The South-Western Mining Syndicate** was registered in 1918 to search for minerals at New Florence, but little appears to have been done. Only a brick built boiler house and some machinery survived from workings of 1918 and 1942. But today the boiler house is in a state of collapse and the Kibble has all but rusted away. However, the steam winch and its boiler still remains, relics of a bygone age.

The ruins of a building can be seen to hold a kibble (a bucket lowered down the shaft). © Copyright Chris Allen and licensed for reuse under Creative Commons Licence

The U-boat campaign in the 2nd World War prompted the Home Ore Department to reopen the mine in 1941. The main shaft was pumped out with an old steam winch the remains of which can still be seen today.

The remains of a horizontal duplex steam winch in situ but derelict.
© Copyright Chris Allen and licensed for reuse under Creative Commons Licence

Work began in earnest in June 1942 when a group of Canadian engineers began driving along the lode. It did not take them long to break into the old workings and were lucky to escape with their lives when water flooded the shaft. After pumping out, the old workings were found to be two and half feet above the floor of the new level extending four-hundred feet west.

The iron ore reached was an average of twenty feet and about three feet wide and extended for about fifty feet. The upper layers were cleared and about twenty tons of ore was brought to the surface. However, flooding remained an insurmountable problem, and according to **The Exmoor Encyclopedia** *local information, led to some fatalities and the abandonment of the project.* However, local man Norman Govier says that there were no fatalities based upon John Rottenbury's thesis, and the testimony of a lady who has lived all her life at Heasley Mill and North Molton. Also, Norman's grandfather worked at the mine during the Canadian engineers time at the mine and he did not report any fatalities.

Chapter 4
THE GREAT GOLD FRAUD OF 1853

THE BRITANNIA MINING COMPANY | THE DUPING OF JOHN MITCHEL | BENJAMIN MASSEY CERTIFICATION | LORD POLTIMORE SMELLS A RAT | MR FLEXMAN'S NUGGET | SIR CHARLES SHARPE KIRKPATRICK CONVICTED OF FRAUD | THE COST-BOOK PRINCIPLE | NO GOLD AT THE BRITANNIA MINE | THE PERKES MACHINE | THE SWINDLE DISCOVERED | THE POLTIMORE COPPER AND GOLD MINING COMPANY | JAMES COOK - AN EXPERIENCED PROFESSIONAL | THE OTHER DIRECTORS | CAPTAIN MOORSOM - CONSULTANT ENGINEER | POSITIVE GOLD REPORTS | DESPERATE MEASURES | THE BERDAN MACHINE TECHNOLOGY | RIPPED OFF? | THE END IS NIGH! | THE POLTIMORE COMPANY FOLDS | AFTERMATH

In his research paper *Gold at North Molton* Dr Peter Claughton of the Department of History, Exeter University wrote:

"The problem at North Molton was not the search for gold itself but the efforts made to promote the schemes, the deception used to substantiate the claims for economic viability, and the blind trust in those promoting and managing the schemes."

Dr Claughton is correct with his analysis. As I researched this unhappy episode in the North Molton mines history it gradually dawned on me that although two companies were said to be involved in the scam, in fact only one was. This was the **Britannia Gold and Copper Mining Company**. This company worked a small mine north of the Bampfylde mine originally called Prince Regent, but renamed the Britannia mine after the name of the company.

Contrary to popular belief the company that worked the Bampfylde mine and was licensed to develop it **The Poltimore Copper and Gold Mining Company** had in fact made a genuine attempt to mine gold based upon some preliminary investigations and a lot of trust in those who advised the company. Their main consultant was Captain W.S. Moorsom. (1804 -1863).

Captain Moorsom was a well known figure of the time, an English soldier who became a civil engineer. With experience of military surveying he made the acquaintance of Robert Stephenson and assisted him in the construction of the London and Birmingham Railway. He was highly respected so it is understandable that the company acted upon his advice and considerable capital expenditure was made even before the viability of gold mining had been thoroughly tested.

As I learned more about The Poltimore Copper and Gold Mining Company the more it became apparent how gullible and incompetent the management team had been. Quite frankly they were very lax and naive about how to run a mine. Little wonder then that they became a prime target for opportunists to take advantage of their ineptitude. It was not long before the company was persuaded to purchase two pieces of very costly equipment, supposedly state-of-the-art, but that simply did not work in practice.

The purchase of the equipment was disastrous for The Poltimore Copper and Gold Mining Company. In the end the company had no option but to wind up its business and cut their losses. Unfortunately, this happened not long after the Britannia Gold and Copper Mining Company swindle had become common news. As a consequence people without any evidence to support their views said that the company had also been complicit with the fraud. By that time the company no longer existed and was in no position to reply to the accusations made against them.

It is time to put matters right and acquit The Poltimore Copper and Gold Mining Company of the scam. Let us take a look at the evidence beginning with the real perpetrators of the swindle the Britannia Gold and Copper Mining Company.

THE BRITANNIA MINING COMPANY SCAM

It is clear from the start that the "Britannia Gold and Copper Mining Company" saw the great Gold Rush of 1849 in California as a great opportunity to make money. So it was not surprising to find that the company begins its prospectus published in May 1852 by describing the success of the recent gold discoveries in California and Australia. Having then wetted the appetites of would-be investors the company then extolled the virtues of local gold.

The company then puts forward two "independant" certifications in order to backup their claim that gold was to be found at the Britannia mining site:

> *The GOLD of this company is produced from gossan and quartz. Several stones, out of a large quantity, equally rich, were promiscuously taken, and the following assays give the result:"*

The first certification entry comes from the pen of John Mitchel.

THE DUPING OF JOHN MITCHEL

This is to certify that I have examined a sample, marked 'No.3 Gold." I find it contains 27.08 per cent of Gold, traces of silver, oxide of Iron and earthly matters", so signed John Mitchel, Assay Office and Laboratory, 23 Hawley Road, Kentish Town, London on 20th December, 1851 in the prospectus.

John Mitchel was certainly a good man to have on your side. He was a recognised expert in the field of gold assaying so a recommendation from him would certainly lend considerable authoritive weight to what the company was promoting. His knowledge on the subject of gold assaying was written down a book published in 1854. Entitled, *Manual of Practical Assaying* published in 1854 he mentions the gold to be found at the Britannia mine twice in the book. First Mitchel says:

> *"Gold from Devonshire and Wales by the Author. The author has received two specimens of gold, one from Wales, and the other from the Britannia Mine, Devon; and found both to be absolutely fine gold.*

John Mitchel also goes on to say:

> *"Native Gold and Aurides of Silver (Native), Au and AuAg n, are found in variously contorted and branched filaments, in scales, in plates, in small irregular masses, in the crevices or on the surface of common ferruginous and other quartz. In Devonshire, at the Britannia Mine, it has occurred in pipes or*

> *veins, and disseminated in a compact hard gossan, one specimen of which was found to contain 27per cent, of fine gold."*

The specimen quoted was clearly the one that Mitchel certified in the Britannia prospectus, but there is a problem with this entry. Mitchel never visited the Britannia mine and the specimen he examined was shipped to him in London. Note what he earlier said earlier namely that he had 'received a specimen' of gold from the Britannia. He was not to know that the sample that he examined had more than likely originated elsewhere and not from the mine. But anyone reading Mitchel's endorsement in the prospectus would not have been aware of the underhanded methods used by the mine owners. As for Mitchel, he did not find out that he had been duped until much later by which time the damage had been done.

BENJAMIN MASSEY CERTIFICATION

The next person who put pen to paper was another gentleman of London - Benjamin Massey. The 1843 **London Post Office Directory** shows him to be a goldsmith and jeweller at 116 Leadenhall address. He was also known as a "Foreign Exchange Officer". In other words, he was not an authority of gold mining as such but a gold dealer who supposedly purchased some rock from the mine and which contained grains of gold. He said:

> *I beg to certify that I purchased 152 ounces of Iron Matrix or Gossan - from the Mine of North Molton, now called the Britannia (promiscuously taken from a large quantity, which, according to my judgement, must have been 2 grains, above the standard, and for which I paid the rate of £3. 17s, 9d, per ounce. I have likewise no hesitation in stating my belief that, from personal inspection of the Mine, there is a large deposit of precious metal to be found there."*

The management of the Britannia mine laid great emphasis on what Benjamin Massey had written for the prospectus. In fact if it had not been for what Massey had said, there was not much else written that might interest a prospective investor. So it is little wonder that the management PR guys milked Benjamin Massey's testimony for all its worth. They went into overdrive to build up a very positive picture of the amount of gold to be found at the site. Following the words of Benjamin Massey, they wrote:

"These consequently would be worth £20,000 to £30,000 per ton, and are equal to the best specimens of Californian produce;". But then the savvy PR guys follow their positive statement with a let-out clause by saying, *"but it is not, of course, to be supposed that the average yield will approach this enormous standard."* However, that did not stop them from providing reassurances by next writing, *" In the future workings, however, it is impossible to say what quantity of the precious metal may be found, but enough may be found, but enough has been ascertained to induce the most sanguine expectations that the result will be highly productive and profitable."* In other words a would-be investor has nothing to lose, as the returns would most definitely be profitable - or say they claimed.

One has to ask, can we be certain that Benjamin Massey was an impartial witness? He was after all just a gold dealer who, like John Mitchel above, who was given some auriferous rock apparently from the mine. In this case the rock contained grains of gold probably barely visible to the naked eye. Further, while it is true that he said that he had visited the mine, we can be sure that although he was shown the gossan dumps, it is unlikely that he was in a position to confirm what he said *"that there is a large deposit of precious metal to be found there"*. In view of the fact that he could not see the gold what he said really was only speculation on his part based upon the assurance of the management team no doubt.

LORD POLTIMORE SMELLS A RAT

Before continuing with the examination of the content of the prospectus and the evident fraudulant details contained therein, it will be pertinent to mention an episode that took place around the same time. Lord Poltimore was approached by the Britannia company requesting that he put his name to the company name.

Lord Poltimore was Augustus Frederick George Warwick Bampfylde, 2nd Baron Poltimore (12 April 1837 - 3 May 1908), a British Liberal politician who served as Treasurer of the Household under William Ewart Gladstone between 1872 and 1874. He married Florence Sara Wilhelmine Brinsley Sheridan, daughter of Richard Brinsley Sheridan, of Frampton Court, in 1858. He held the mineral rights to the land and his name would give the company considerable creditability if he could be persuaded to sponsor them with his name.

What transpired next is to be found in the book published by the North Molton History society where Peter Stucley writes:

> "On the subject of mining, I think it is worth mentioning here the excitement that surrounded the prospecting for gold in the 1850s and Lord Poltimore's scepticism. There was a body of speculators who, enticed by a curious reference in the Church wardens' accounts of 1773-1774, sought leave from Lord Poltimore to mine on his land... On 23rd May 1852 there was a gold strike and, as a result of all the following publicity, the Mining Journal was extremely optimistic."

The prospectus of the Britannia Gold and Copper Mining Company had been published in May 1852 and in it the company claimed that they discovered a vein of pure gold a quarter inch thick. This clearly was the gold strike of which Peter was referring. To try to persuade Lord Poltomore to add his name to the company, they then made a fatal error. Peter Stucley noted:

> "A nugget was found and shown to Lord Poltimore who remained sceptical and refused to lend his name to the company's activities."

It was evident that Lord Poltiomore knew the history of the nugget. It knew that it had not been found recently but had been discovered in 1810 under dubious circumstances.

MR FLEXMAN'S NUGGET

In 1819 *The Journal of Science and the Arts* reported a remarkable find. Apparently, a gold nugget was found at the Prince Regent Mine (Britannia mine) by a gentleman called Mr Flexman. The journal wrote:

> "Native English gold has been found lately in Devonshire by Mr Flexman, of South Molton. It occurs in the refuse of the Prince Regent mine, in the parish of North Molton; the mine was discovered in 1810, and worked for copper, but was discontinued in 1818. The refuse is a ferruginous fragmented quartz rock, and contains gold in imbedded grains and plates."

The observant reader may be wondering what the aforementioned nugget has to do with the Britannia Gold and Copper Mining Company. It just so happens that when the members of the company approached Lord Poltimore to ask for his backing by allowing them use his name, they showed him a nugget in order to impress him. Lord Poltimore was not impressed and the fact that it was presented in the first place made him suspicious. Why? If you take a look at the prospectus for the Britannia Gold and Copper Mining Company, who should appear as a member of the 'Committee of Management'? Yes you guessed it! William Flexman.

The prospectus goes further and identifies him as *the gentleman who first discovered the Gold deposit at the Britannia mine..* Do you smell a rat? Lord Poltimore did. Obviously, the nugget shown to him was the same as the one that Mr Flexman had supposedly found in 1810. Lord Poltimore was not so gullible to be misled by such a ruse.

Although Lord Poltimore was suspicious of what was being presented as evidence of gold at the Britannia mine, the management team of the company did manage to convince him to allow them to establish mine workings on the site of the former Prince Regent mine. They never had his personal backing though with respect to his name, unlike the company that was to establish mining operations further south at the Bampfylde mine - **The Poltimore Copper and Gold Mining Company.**

SIR CHARLES SHARPE KIRKPATRICK CONVICTED OF FRAUD

Foiled by not being able to use Lord Poltimore's name they found someone else who had a title and who was more than willing lend his name to the prospectus - for a price. His name was Sir Charles Sharpe Kirkpatrick (1811-1867) 6th Baronet of Closeburn in the County of Dumfries. His price was to be become a member of the Britannia Gold and Copper Mining Company management team. He was well known as a gold speculator and was at the time the primary share holder of Cwmheisian Gold Mining Company that worked the Ystrad Einion mine (Cwmheisian mine), Ceridigion in Wales for gold.

A letter published *The Sydney Morning Herald* of 25 July 1854 is worthy of mention here. Written by George Windsor Earl, he says the following:

> "Sir, In your article of to-day on the case of Wyld v. Calvert,

which was argued before the Court of West- minster, in April last, you have made some allusions to the alleged gold discoveries in England, which may possibly have been based on the information I considered it my duty to communicate to you respecting the failure of the Cwmheisian mines of North Wales, of which such high anticipations had been formed."

Mr Earl now elaborates and describes what happened at the Cwmheisian mines, and interestingly we find that a Mr Berdan is involved in the plot, of whom we shall learn more about later in this chapter. Mr Earl continues with his letter:

"You will therefore oblige me by inserting the following statement, which contains all the particulars I am at present possessed of. In my letter commenting on the article in Chambers' Journal, on the gold discoveries in England, which appeared in your issue of May 25th, I mentioned that Mr. Calvert had been introduced to Mr. Berdan and his agents by the Directors of the Cwmheisian Mining Company (Sir C. Kirkpatrick, Bart., Mr. T. A. Readwin, F.G.S., and Mr. Dickinson Brunton, C.E.) as one of the parties appointed by them to superintend the experiments in their ores with his machines, and that he (Mr. Calvert) was present on the occasion in which the Bank authorities and a number of scientific gentlemen witnessed the trials. The results of these and subsequent experi- ments were so favourable, giving proceeds varying from 1 oz. 1 dwt. 15 grs. to 5 oz. 5 dwt. 17 grs. of gold to the ton, that the Directors were induced to commence working their mines on a large scale. At first the operations seem to have yielded an average of about an ounce of gold to the ton, but they suddenly ceased to yield even the slightest trace; and Dr. Henry, an eminent assayer, who was sent down to ascertain the cause, was unable to detect any gold in the ore."

Surprise, surprise! There was no gold at the Cwmheisian mines, it had all been an elaborate fraud. Mr Earl writes:

"We may rest assured that the whole affair will be thoroughly sifted, and if the ores have been, what is here technically called "peppered," the delinquent will be brought to justice, for large

sums of money must have changed hands in consequence of the alleged discoveries."

The delinquent was indeed brought to justice. As announced in the *London Gazette*, December 18, 1857, Sir C. Kirkpatrick and John Dickinson Brunton both directors of the Cwmheisian Mining Company were to be sued and to appear before Mr Commissioner Philips on 2nd January 1858 to answer the charges level against them regarding their part in the fraudulent practices carried out by the company.

Sir Charles Kirkpatrick, one of the directors of the Britannia Gold and Copper Mining Company being sued for a similar gold mining swindle in 1858 as a major shareholder of the Cwmheisian Mining Company in Wales

I have not been able to ascertain whether Sir C. Kirkpatrick and John Dickinson Brunton were found guilty or not but for the purposes of this book it is enough to learn that Sir Kirkpatrick, a director of the Britannia Gold and Copper Mining Company, had also been involved in an almost identical scandal, a gold mining operation in North Wales.

THE COST-BOOK PRINCIPLE

From the start, some if not all the directors of the Britannia Gold and Copper Mining Company had set out to cash in on the Californian gold fever that was sweeping the world, and saw the Prince Regent mine as the chicken that laid the golden egg. Although they knew there was no gold to be found there, they also knew that it would take a long time before any swindle could be discovered. In the meantime, they could fleece unsuspecting speculators of their money and then escape without being held to account. Surely, one may ask, how could they get away with such a swindle? The secret is simple. The Britannia Gold and Copper Mining Company offered 36,000 shares of a £1 each based upon what is called the Cost-book principle.

THE
BRITANNIA
GOLD
AND COPPER MINING COMPANY,
NORTH MOLTON, COUNTY DEVON.

CONDUCTED ON THE "COST-BOOK" PRINCIPLE.

Thirty-Six Thousand Parts or Shares of £1. each,
IN CERTIFICATES TO BEARER.

What the cost book is about is described in the book *The Gold Companies and the Cost-Book System* by Joseph Napier Higgins published in 1853. In the book we read:

> *"The history of a Cost-book company proper is obvious enough. A few men have discovered a mine, agree to join together as partners to work it. They further agree to divide the profits at short intervals, and it is understood that each has an equal right to control the adventure, and an interest in it commensurate with the amount of money which has been paid into the common fund."*

At first glance the Cost-book principle may seem like a sensible solution for obtaining funding while shareholders received a dividend proportionate with the amount of money that had been invested. However, there is a fundamental flaw with the system. According to *The Cost Book Its Principles and Practice* by Thomas Tapping, the Cost-book principle is a voluntary commercial usage in the nature of an ordinary common-law partnership applicable to the working of mines by an association of adventurers.

The words of Thomas Tapping is echoed by Jospeh Higgins aforementioned:

> *"The origin and history of the Cost-book system are easily given It needs little research to show that it is merely a local custom about which have grown up many peculiar customary laws that have no force or application beyond the locality where they are recognised and sanctioned by the statute or common law of the land".*

The major flaw of the Cost-book principle therefore is that it is not a legal contract but is a common-law partnership and as such it is easily open to abuse. Control and the way the money is utilised is controlled by the directors and not the shareholders. It is true that the shareholders had a right to meet bi-monthly or monthly, to call for a balance sheet from the directors, and if necessary they had the right to depose the directors and appoint new ones, but that is all. Consequently, if the directors doctored the books and kept the shareholders in the dark as to what they were doing, there was little the shareholders could do if later they found out they were victims of a swindle.

The Cost-book principle was the perfect system for anyone who wants to implement a scam, and the Britannia Gold and Copper Mining Company milked it for all its worth. They knew that in the end, it would be very unlikely that they would be sued once their despicable deeds had been discovered as author Joseph Higgins aforementioned above notes when describing the flaws of the system.

> *"The truth appears to be that the shareholders having subscribed their money, thenceforth left its disposal completely at the discretion of the directors, and, under the circumstances - taking into account their relation to law - it is difficult to suggest what legal process or proceeding in Equity could be resorted to for the purpose of bringing the directors successfully to account for the improper application for the funds in the company. In the event of the insolvency of the concern, there is equal if not greater difficulty as to its winding up."*

NO GOLD AT THE BRITANNIA MINE

The nugget supposedly found at the Britannia mine by Mr Flexman originated elsewhere, as did the samples that were sent for analysis from the mine, which were evidently 'seeded. Reports of gold sometimes reported in the area were not found at the Britannia mine but at the Poltimore site further south, as Peter Claughton notes.

> "The name Prince Regent, and the gold discoveries, have hitherto been linked with the Britannia Mine site but contemporary references to gold being found in the 'rich' copper mine clearly identify it as the Bampfylde Mine site."

It is clear that any search for gold at the Britannia mine was doomed to failure, but the directors kept up the charade for as long as they could while they syphoned of funds to fill their pockets. When ore was sent for testing, and came back negative, the directors argued that they needed a better system to extract the gold. So they invested in a Perkes machine, probably exaggerating the cost while lining their pockets at the same time.

THE PERKES MACHINE

The owners of the Britannia mine invested in a Perkes machine. How much of the money went to actually buy the machine and how much went into the pockets of the directors is not known. Knowing now the fraudulent backgrounds of the directors it is not unreasonable to assume that the books were doctored to show the total cost of the device while hiding what the directors took and the real cost of the machine.

A Perkes machine manufactured by Barratt, Exall and Andrews was installed at the Britannia mine at great expense. It failed to process any gold.

The Perkes machine was a cast iron pan, six feet in diameter and three feet six inches high, in which five heavy cast iron cones revolved, driven by a central vertical shaft. It was manufactured by a reputable company namely Barrett, Exall & Andrews of Katesgrove Iron Works, Reading. Founded in 1817/18 and, employing up to 360 people their manufacturing facilities occupied a 12 acre plot. They produced agricultural machinery and portable/fixed engines. There is no suggestion that they were involved in the swindle. As for the machine itself Mercury was used to create an amalgam, a mixture of gold and mercury while the rock debris was washed away by water.

In 1854 the first 50 tons of ore was processed by the Perkes machine and it took over four weeks of day and night work before the results of their labours could be determined. Needless to say the outcome was pitiful with only 14 grains of gold per ton of ore extracted as the *Mining Journal* of 1860 relates:

> *Gold extraction proceeding but slowly, barrel amalgamation was substituted, the ore being ground dry in the mills, which worked very effectually, reducing the gossan to an impalpable powder ...... This arrangement was superseded by Perkes' machine, a cast iron pan, six feet diameter and three feet six inch high, in which five heavy cast iron cones revolved, worked by a central vertical shaft. Numerous working trials were made ...... one upon 50 tons of auriferous gossan. The time occupied in reduction and amalgamation was four weeks of day and night work, and the final results were a loss, by*

disintegration, of 50 per cent of the mercury employed, and a ultimate yield of 1½ ounces of gold, or 14 grains per ton of ore. Every attempt to extract gold from North Devon ores remuneratively by this machine proved a failure.

THE SWINDLE DISCOVERED

The Perkes machine is only as good as what you put in. So it is not going to produce any gold if the ore that is supplied contains none. For weeks and then months every attempt to extract gold by the Perkes machine or any other process proved a failure. Had it not been for news that at the Poltimore mine to the south where an assay of gossan from the western side of the river gave a return of 11 ounces per ton of gold, the swindle would have been found much earlier, but the news staved of an immediate revolt by the shareholders of the Britannia Mining Company as they hoped to see the same success at the Britannia mine.

It would only be a matter of time before the shareholders who were expecting great dividends for their investment would wise up to the fact that there was no gold. Where was the gold that had been promised? Where were the nuggets like the one found by Mr Flexman? These and other questions were now being asked and rumour of a massive fraud began to be circulated. *The Devonshire Director* journal of 1856 says it all. "*a great fraud was got up in 1853, and a sham gold mine foisted on the public.*"

With the swindle now discovered the directors of the Britannia Gold and Copper Mining Company having made their money out of the company saw no need to carry on the charade any longer and set in motion the closure of the company. *The Mining Journal* of 19th September 1857 describes the closure.

> "*Special general meeting of the Britannia Gold Mining Co called for October 28, for winding up affairs. First Gold Mining Co in England. Large amount on machinery by Capt Moorsome R.E. Many thousands in mining and machinery spent, a perfect failure.*"

The dishonest Britannia Gold Mining Company management lamented the closure of their enterprise. Through the same journal, they declared that they had done their best but very little gold had been found.

"Every exertion made but a commercial failure. Machinery erected at own cost of a member £1000. Lodes are auriferous proved in some places better than others but average AU insufficient to pay costs of extraction......Shareholder suggested payment ex gratia to men injured through working with quicksilver. All liabilities discharged."

So it was that, knowing that gold was never to be found in any conceivable commercial levels, the directors of the Britannia Gold and Copper Mining Company got away with the swindle knowing that they could not be sued under the Cost-book principle - all liabilities had been discharged. How much of the £36,000 that had been raised by shares lined their pockets, we shall never know.

THE POLTIMORE COPPER AND GOLD MINING COMPANY

It should be said from the start that despite popular belief or rumours the Poltimore Copper and Gold Mining Company was never involved in the swindle at the Britannia mine nor was it involved in a swindle of its own. If anything the company was the victim of the scam, of which I shall relate later.

The prospectus of the Poltimore company was published in October 1852 by which time the Britannia mine was fully operational with 50 tons of gossan having been raised and a six ton sample despatched to Messrs Johnson for a large scale assay. By all accounts the prospectus was less sensational and more factual. Unfortunately, I have not been able to obtain the prospectus of this company but I do have an abstract that was published in the *Devonport Journal*, 11th November 1852 from which I have gleaned some useful information.

Like the Britannia Gold and Copper Mining Company the Poltimore company was established on the Cash-book principle, but unlike their northern rivals, no swindle was planned or implemented. In fact it offered guarantees. 50,000 shares of £1 each were issued, with a minimum guaranteed return of five per cent showing that the company was in business for the long term. No such guarantee was made by the Britannia Gold and Copper Mining Company.

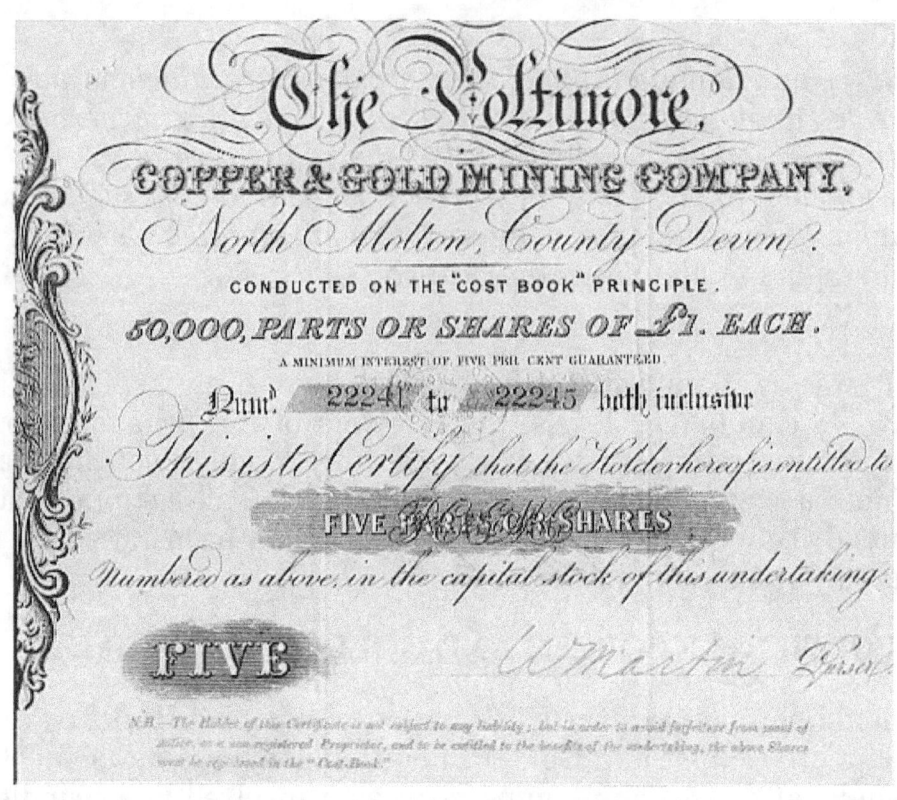

Another factor to take into consideration about this company is that it carried the good name of Lord Poltimore. He must have considered the company a business worthy of his trust. The Bampfylde mine was renamed the Poltimore mine as a result.

JAMES COOK - AN EXPERIENCED PROFESSIONAL

The prospectus did not just focus on just the gold that might be present at the Poltimore mine, but it also described its copper mining potential too.

> "The sections show that the Poltimore mine is no virgin mine, and the ticketings that the produce of copper ore is no mean standard..... The average yield of 20 ticketing was 15 ½ per cent, while the average yield of copper ore in the United Kingdom is from 7 to 8 per cent. Tributors are already at work at 13s 4d in £1, and there are 100 fathoms of good orey tribute ground now ready....Throughout the vast refuse heaps which now remain are great quantities of rich copper bearing stones, which in the west would have been sent to stamps, and of themselves would indicate the presence of a good copper mine."

With a long history of copper mining at the Poltimore mine and evidence that more could be excavated in commercial quantities it is of considerable interest that we learn that one of the directors of the company was James Cook. He was already involved with a successful copper mining enterprise - the **Devon and Cornwall United Copper Mines**

James Cook was the chairman of the said company which worked the George & Charlotte Mine and the William & Mary Mine, close to Morwellham Quay that lies on the River Tamar, 35 km inland from Plymouth. These mines were rich in childrenite, the finest in the UK, but it is copper that was the main source of income for the company. According to the 1856 **Post Office Directory** output from these mines between 1852 and 1869 was 16,595 tons copper ore and 110 tons of pyrite, meaning childrenite. The company also operated on the Costbook system that consisted of only 1,024 shares, most of which James Cook owned.

THE OTHER DIRECTORS

With the exception of James Cook the other directors of the committee of management of the Poltimore company were primarily amateurs, speculators mainly from London. They were Frederick Chase (Tiverton), Charles Heneage (London), Thomas Inglis Hampton (London), Benjamin Massey (London), Richard Martin (London), Henry Mogford (London) and Henry William Taylor (North Brixton). There were no locals as part of the management committee and the team's access to local knowledge was minimal.

At this juncture it should be noted that one of the directors of the Poltimore Copper and Gold Mining Company was none other than Benjamin Massey, the same person and goldsmith who was described above and who provided certification that the ore sent to him by the Britannia company contained gold. Could it be that he discovered that the auriferous ore had not originated at the Britannia mine but had actually been excavated from the Poltimore mine and saw an opportunity to cash in on this insider knowledge? Why else would he become a director of the Poltimore company and not the Britannia company?

The only real expert in mining operations was therefore was only James Cook and he was in the enviable position of being able to bring his experience and expertise of copper mining to the management team. But the problem was that he was away far too often dealing with his other mines, that his expertise could not be called upon on a day to day basis. It was left to the other directors of the committee of management to rely on advisers to guide them but here too their main advisor had never done been involved in copper or gold mining before. Hence, from the start of the enterprise, the directors were already handicapped by their lack of experience and knowledge which was later to prove their undoing.

CAPTAIN MOORSOM - CONSULTANT ENGINEER

The main advisor to the Poltimore Copper and Gold Mining Company was consulting engineer Captain W. S. Moorsom, who as it happens was also the consultant advisor of the Britannia company. His title of Captain was not because he was the captain of the mine but rather he was formerly a captain of the 52nd light infantry in Nova Scotia. It was a status symbol that he preferred to be known by.

Here is a little background information to put things into prospective. After his education at the Royal Military College, Sandhurst, William Scarth Moorsom was commissioned an ensign in the British army on 22 March 1821, and joined the 69th Regiment on 7 Nov. 1822. He went on half pay in the Cameron Highlanders before being raised on 8 April 1826 to captain in the 52nd Regiment, which he joined in Nova Scotia the following August.

Before Captain Moorsom returned to England with his regiment in 1831, he had explored most of mainland Nova Scotia, mapped Halifax harbour, commanded the small army detachment in Prince Edward Island, and served as acting deputy quartermaster general for the Nova Scotia command from November 1830 to September 1831.

On 2 March 1832 Captain Moorsom sold his army commission to care for his father, hoping to return to Nova Scotia shortly and settle on the land he had purchased in Hants County. Things did not work out as he had planned an instead he sold the land in 1835 and through his brother who was director of London and Birmingham Railway Company he became involved in the railway industry where he found his true vocation and excelled. His survey of a very difficult section of country, crossing the valley of the Ouse, even attracted the notice of Robert Stephenson, the famous locomotive designer and bridge builder and son of George Stephenson, the famed locomotive builder and railway engineer.

During 1838 to 1834 Captain Moorsom visited and studied every railway and canal working or in progress of construction in England. He became a highly respected engineer and was employed in laying out many railway systems in England and Ireland. In 1845 he received a Telford medal for the first practical application, in the construction of the cast-iron viaduct over the Avon at Tewkesbury, of the method of sinking iron caissons by their own weight in a river-bed, pumping out the interiors, and filling with concrete to form the piers.

There can be no doubt that Captain Moorsom was a brilliant engineer but the cessation of railway enterprise from 1852 necessitated him to look for work elsewhere. It was then that he took on a position as consultant engineer for both the Britannia and Poltimore mining companies where his engineering construction skills were much in demand.

Although a brilliant engineer, Captain Moorsom had no experience in gold mining so that when he was called upon to send samples of ore to London for anaylsis on behalf of the Britannia mine, he would not have known what he was sending. But the Britannia Gold and Copper Mining Company got what they wanted, a man of distinction to add to their portfolio of experts, and got him to write a report promoting the mine even though he had little knowledge about it.

Through his acquaintance with Sir Charles Sharpe Kirkpatrick, Captain Moorsom learned about Wilson Berdan and his invention the Berdan machine. As an engineer he could appreciate its inner workings and mechanism for assaying gold. As a result he became an enthusiastic advocate for its use for the mining operations of the Poltimore company. Naive, he was not to know that this would prove disastrous for the company and that his uninformed advice would be responsible for its sudden demise.

POSITIVE GOLD REPORTS

While the Britannia company was going through the motions of finding gold at their mine with negative results, the first shipment of ore by the Poltimore Copper and Gold Mining Company proved positive. If you may recall in a previous chapter, the cutter Albatross was to deliver 100 tons of auriferous gossan from the Poltimore mine to Liverpool. This the cutter did and the tests on it proved positive as the **North Devon Journal** of 17th February 1853 acknowledges. *"The initial results in Liverpool were varied, one of 6 dwt per ton and another considerably higher at 10 ounces per ton.*

This one good result encouraged the directors to continue mining at the Poltimore mine and it gave a lifeline to the directors of the Britannia company who used the good result to persuade their shareholders that they had had a run of bad luck. In his paper *Gold at North Molton - and*

the surviving evidence to be found on the Bampfylde Mine site, Devon.
, Peter Claughton remarks:

> *" In the meantime an assay of gossan from the western side of the river at the Poltimore gave a return of 11 ounces per ton, encouraging both companies to continue with plans to raise gossan in large quantities."*

Keep in mind that the encouraging results came from ore excavated on the west site of the Poltimore mine. This is important as we shall discover later in this book.

DESPERATE MEASURES

It is true that the Poltimore company had found some gold but the mixed results from the Liverpool assay and other assays was a cause for concern. After twelve months of working only about £180 profit had been made with barely fifty ounces of gold produced. The cost of extraction, transportation and testing was proving far too expensive. This was admitted by one of the members of the management committee, Henry Mogford. How he got funding to go to Australia is open to question but during a lecture delivered before the Society of Arts by Charles F. Stansbury in Australia it is reported that he said the following:

> *"Mr. Mogford, of the Poltimore Mining Company, said, after the observations of the last speaker, he felt bound to make a few remarks. The Poltimore mines, in Devonshire, had been at work for the last twelve months, and forty tons of ore had been sent to Messrs. Rawlings and Watson, of St, Helens, near Liverpool, for reduction. The cost had been : In bringing the ore to grass 3s. per ton, freight 17s, and reduction 30s., whilst the produce had been about fifty ounces of gold. Since that time, 120 tons had been sent to the Messrs. Rawlings for reduction, who, after deducting overy charge, had sent them a cheque for £170 or £180, as the profit."*

What Henry Mogford said in reply was reported in the *The Sydney Morning Herald Monday 6 March 1854* and clearly £180 was not much to show for the extraction and processing of 120 tons of ore. While it was true that gold was present in the gossam of the mine it was just too costly to be sent away for processing. There had to be another solution, one where the gold could be extracted insitu. It was then that Henry Mogford told the attendees of the lecture that the company had decided to invest in the latest technology that had only just become available. He said:

> *"They had since entered into an engagement with Mr. Berdan to erect machines at the mines, where there was an almost unlimited supply of gossan, from which they expected to obtain gold at an expense not exceeding 10s, per ton. They had tested the machines at Mr. Berdan's, in the presence of Mr. John Wilson, of the firm of Rawlings and Watson, when the brown gossan, of which there were specimens in that room, produced 13 dwts of gold to the ton, and the red gossan 32 dwts and Mr. Wilson expressed himself perfectly satisfied with the working the reduction which, under the system of his firm, would have cost 30s. per ton, being made at a comparatively trifling cost. He believed that in twelve months they would be in a condition to prove that gold could be most profitably produced in England.*

Henry Mogford was as good as his word. The new technology that he referred to was the Berdan machine. It was a device designed in New York in 1852 by Wilson Berdan, after one of his engineers had visited California to investigate the crushing of quartz. The engineer reported that all processes except assaying with mortar and pestle lost as much as 30 to 50% of the gold. Like all inventors, Wilson Berdan saw a niche and a great opportunity to make money. So he set out to duplicate the assaying procedure on a large scale through mechanical means and built a machine named after himself.

Wilson Berdan first exhibited his crushing and recovery machine in 1852 and before long orders were pouring in. By 1953 the machine was increasingly being used in Calfornia and North Carolina proving to be successful, so says the journal *Gold Diggers Advocate* 27 May 1853.

There is a wisp of evidence that when Wilson Berdan carried out his demonstrations that he seeded the ore being crushed to get the remarkable results that stunned his potential clients. He set up free testing facilities in London that proved very popular and the results were amazing....to amazing to be true. This is evident from what is written in the book *Gold Mines of Meninydd* by George Hall. This is what he has to say:

> *Free testing facilities proved so popular that by mid-December the plant had opened every day, instead of Wednesdays only, as originally intended and Berdan was astonished at the number of British localities which showed high values - as many as two hundred were eventual recorded. Berdan and his staff would allow no one to interfere with either the grinding process or assaying of the resultant amalgan..."*

I think you can see what was going on here. To get such incredible results from auiferous rock from two hundred British localities is pushing credulity to the limit. With Wilson Berdan and his staff not allowing anyone to interfere with the process of the machine I think you can work the rest out for yourself. His slight of hand technique of seeding the amalgam was proving very effective and British clients shown to be very gullible. One of these the Poltimore Copper and Gold Mining Company was to find this out the hard way and to their cost and ultimate demise.

THE BERDAN MACHINE TECHNOLOGY

A Berdan pan is essentially a heavy duty grinding machine. It is a large circular pan (basin) about three and half feet in diameter set at an angle. As the pan revolves, one or more heavy iron ball rotate in the lower part, grinding rock or minerals to a fine powder. During the process mercury is poured in to make an amalgam of gold and mercury while the rock debris is washed away. The pan and mechanisms involved are made of iron because it is one of the few metals that is unaffected by the mercury.

The angled pan allows the lighter particles to be slopped over the side by water, while the gold, as it is released and scoured, amalgamates with the mercury at the bottom of the pan. Then the amalgam is removed so that the gold can be extracted from the mixture in a separate process. Thus, the machine crushes, washes, and amalgamates all in the same operation.

A Berdan machine in action

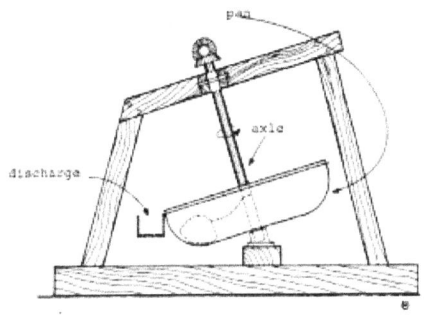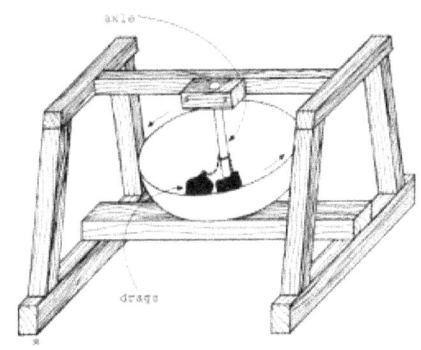

RIPPED OFF?

One of the first installations of Berdan machines in the United Kingdom was at the Cwmhesian Mine in Wales, the same one that Sir Charles Sharpe Kirkpatrick was a shareholder, and who also was a director of the Britannia company. The *Perth Gazette and Independent Journal of Politics and News* of Friday 18 August 1854 reported the installation in a very positive light:

> *"Two of Berdan's double-pan machines have been set up at the Cwmhesian Mine in Wales, which we are informed have worked exceedingly satisfactorily, and produced from four tons of very hard quartz in two pans, on Tuesday, 44 ozs. of nearly pure gold in nine hours, with every prospect of increased returns, as the basins get into more free working order, and the "workmen better ini tiated in their employment, it being estimated that the four basins will reduce 22 tons per day."*

We now know that just as the Britannia mine perpetrated a swindle so did the **Cwmheisian Mining Company**. It was evident that the aforementioned positive report was fraudulent and as we shall see later, Wilson Berdan is likely to have been party to the fraud too. In any case, the person who introduced Berdan machines to the management of the Potlimore mine was none other than Sir Charles Sharpe Kirkpatrick.

On specification, the Berdan machine seemed to be the perfect solution to resolving the high cost of extracting gold from the Poltimore mine. Captain Moorsom as an engineer could see that the mechanism used by the machine was based upon sound scientific principles. It all made sense because by processing the ore on site it was bound to be make the chances of making the mine a profitable concern. His own systems had proved too slow and inefficient so here was a solution that solved his problem. Little wonder Moorsom was very enthusiastic to use the Berdan machine.

Captain Moorsom may have been a great railway engineer but when it came to mine workings he was clearly out of his depth. Besides the slow ore processing machines he created, he also made an expensive mistake when he tried to build a 50 foot water-wheel on the western side of the river. The idea was to use it to pump the main shaft (Engine Shaft) located within the main mine workings on the east side of the river in order to power the Berdan machines. As Peter Claughton says:

> *"it was a further example of Moorsom's ill conceived ideas. There was insufficient water on the western side to power a wheel of that size and it was never completed."*

We now come to an extraordinary revelation. Remember that we are talking about new and expensive technology that still had not been entirely proven, so it comes quite a surprise to learn that not one but two Berdan machines were installed at the Poltimore mine at great expense. The only explanation for this madness is perhaps that by buying two machines this was the quickest and best way to get profitable returns.

Who did the convincing? Sir Charles Kirkpatrick quickly comes to mind. If your recall he had two Berdan machines installed at his Cwmheisian mine and reports concerning them were incredible, too incredible to be true. It is quite likely therefore that Sir Charles Kirkpatrick had a lucrative financial arrangement with Wilson Burdan who made it worth his while to promote his products.

The two Burdan machines were installed at the Poltimore mine in May 1854 and high hopes were entertained by all. The machines had cost so much money that there was little money left in the bank. The machines needed to get immediate results in order for the business to continue. It was by now a make or break situation.

Disaster! Working day and night, each Berdan machine processing three tons of gossan per day as against the ten tons per day projected over a long period, no gold was produced. Similarly, a 48 ton batch of ore treated in Liverpool produced no gold either. All the machines succeeded to do was to pollute the land with quicksilver (mercury) and to injure workers with a platter of chronic illnesses such as sensory impairment (vision, hearing, speech), disturbed sensation and a lack of coordination.

As disaster loomed for the Poltimore company, questions were being asked as to why the Berdan's machines had been successful when tried prior to the company placing their order. The finger of blame was pointed at Berdan, with the suggestion that he had introduced the gold into the gossan that was tested. Unfortunately Wilson Berdan had left the country and was unable, or unwilling, to answer the charges. Could Mr Berdan had done such a thing? It just so happened that when the swindle at the Cwmheisian Mining Company was being investigated, suspician also was directed towards him too.

A gullable George Windsor Earl in his letter published *The Sydney Morning Herald* of 25 July 1854 could not believe that Wilson Berdan had seeded the ore during testing of his equipment. He wrote:

> *"There can be no doubt but that Mr. Berdan must have made a large sum by the sale of his machines, but it is scarcely probable that he would countenance measures which would sooner or later entail the loss of a high scientific reputation, and all for the sake of a mere temporary addition to the wealth he had already accumulated from other sources."*

There is a saying that lighting does not strike twice. I am convinced that Wilson Berdan had been complicit in the frauds perpetrated at the Britannia and Cwmheisian mines. What do think?

That there **IS** gold to be found at the Poltimore mine as the earler reasonable returns from the Liverpool shipments had proved, as well as other evidence that I will offer later in the book, of this I am certain. The problem for the Poltimore company was that the management team had not grasped the patchy nature of deposition of free gold scattered on the site. However, I can now, with reasonable certainty, point to where the best returns of gold can be found. All will be revealed later in the book.

THE END IS NIGH!

The Poltimore Copper and Gold Mining Company was by now in dire straits financially. I have been able to obtain the last minutes of the Poltimore company before it had to finally close. It was published in the *Devonport Journal:*

"Thursday 18th January 1855. *(Devonport Journal)*

> *At the meeting on Wednesday (Mr. C. Heneage in the chair,) the accounts showed a balance in favour of the mine of £3, 558. 11s 9d. The report of the committee of investigation, and the reply of the directors, was ordered to lay on the table. A report from the committee of management, recommending the vigorous prosecution of the mines for copper was, unanimously adopted."*

The company management had decided to concentrate on copper mining instead of gold it the hope that this would resolve their financial difficulties. But it was too little, too late. The next meeting in May sees the fortunes of the company no better off in fact things had got decidedly worse. You can almost feel the pain that punctured the scene.

> **"Thursday 3rd May 1855.** *(Devonport Journal)*
> "At the meeting on Wednesday (Mr. C. Heneage in the chair,) the accounts showed a balance in favour of the mine of £1,391. 7s. 10d. A call of 1-shilling per share was made. The committee of management and purser were re-elected, and a special meeting was called for the 13th of June, to ascertain the number who had responded to the call; as in the event of its not being fully paid up the money would be returned. The proceedings terminated with a vote of thanks to the chairman."

In a last ditch effort to salvage the company, the committee of management now appealed to the shareholders for more financial backing and requested that they gave one shilling for each share they owned to help the company out of their difficulties. The appeal fell on deaf ears.

> **"Thursday 21st June 1855.** *(Devonport Journal)*
> At the meeting on Wednesday (Mr. C. Heneage in the chair,) resolutions were passed extending the time for payment to the 30 inst., the holders of 18,575 shares having already responded to it. In the event of all the calls not being paid by the time named, the committee were authorised to wind up the concern, it being agreed that the whole call shall be returned. In the meantime the necessary expenses of working the mine are to be defayed out of the assets in hand, which it was expected to meet all liabilities. The proceedings terminated with the usual complimentary votes."

THE POLTIMORE COMPANY FOLDS

Unfortunately, the call for further investment by the shareholders did not materialise and the company had no choice but to cease operations. The company had genuinely searched for gold but had done so in such a haphazard way that they failed to appreciate the patchy nature of deposition of free gold in the mining area.

The lack of expertise within the management team and one has to say, there incompetence, all played a part in the downfall of the company. So too was the fact that one of their number was a rogue who sang from a different 'song book' but who later would get his just rewards for his fraudulent activities. The management also relied heavily on the advice of their consulting engineer who although a well respected engineer of railways had no experience in working mines. Captain Moorsom had made some very costly mistakes for the company.

What really drove the nails in the coffin was the company purchasing two expensive untried Berdan machines. This was truly an act of supreme folly and they were not to know that Wilson Berdan was a man who could not be trusted. He had fooled them with a demonstration of his equipment with seeded rock, while the traitor in their midst walked away with commission for introducing the Berdan machines to the company.

It is clear from my investigations that The Poltimore Copper and Gold Mining Company was not involved in the swindle that had taken place at the Britannia mine. We can see from the minutes of the committee of management, as the company was in its death throes, that they did not want to wind up the company. Unfortunately, until the publication of this book, the company became subject to all kinds of whispers and rumours so that in the end it became tarnished with the same brush as the real swindlers, The Britannia Gold and Copper Mining Company. The truth was that there was gold at the Balfylde mine, it was just that the management of the mining company was incompetent.

AFTERMATH

The Poltimore Copper and Gold Mining Company had operated the mine for three years, from 1852 to 1855 before mining operations at the mine ceased. The mine was sold to another company the name of which I have not been able to track down. It is only known that the mine was worked for copper between 1856 and 1864. No longer called the Poltimore mine, the mine returned to its former name, the Bampfylde mine. The following is a summary of the companies involved in the mining operations at the mine until its final closure in 1887.

- Poltimore Copper and Gold Mining Company (1852 - 1855)
- Name of Company unknown who mined for copper (1856 - 1864)
- New Bampfylde Copper Mining Company (1865 - 1872)
- Bampfylde Copper Mining Company (1874 - 1882)
- North Molton Mining Company (1883 - 1884)
- North Devon Mining Company (1885 - 1887)

With the final closure of the Bampfylde mine in 1887, gradually, the area became overgrown as nature gained back what was once lost to it. Mine shafts were filled in, and the buildings that remained crumbled into ruins. Apart from large piles of gossan littering the place and the ruins of the crusher house that is being restored by Norman, a visitor of the site would never have known that it had been a hive of activity where hundreds of people had been employed, and that a major business enterprise had operated from it. Only a few people now know what it had once been like, thanks to the tales that were passed on to them by their grand parents and great grand parents describing those days.

Chapter 5
I VISIT THE BAMPFYLDE MINE

LOCAL KNOWLEDGE MAKES ALL THE DIFFERENCE | AN INVITATION I COULD NOT REFUSE | THE ADVENTURE BEGINS | I MEET THE YOUNGEST SON OF SIR HUGH STUCLEY | ACCESS TO BAMPFYLDE MINE | THE CRUSHER HOUSE | NORMAN SHOWS ME SOME GOLD | THE ENGINE SHAFT | WHERE THE BERDAN MACHINES WERE LOCATED | END OF FIRST VISIT

Although I had gathered quite a lot of information about the mines around North Molton it was clear there was gaps in the information that was missing. What was certain was there had been an enormous scam that taken place in 1853 at one of the North Molton mines where speculators had sought to capitalise on the almost mass hysteria surrounding the Californian Gold Rush of 1849.

I was left with a conundrum. OK! A fraud may have been perpetrated but what about the rock that contained gold that was on display at the South Molton Museum, the one that I had held in the palm of my hand the week before? Had that been planted as part of the scam too? What about the other reports of finding gold in the mines long before the fraudsters came on the scene? Were they also fraudulent? I did not believe so. There was too much evidence, real and circumstantial, that suggested that gold had been found at the Bampfylde mine.

It was clear that I needed help to sort out the dross from the real stuff, someone who had knowledge that I did not possess. Fortunately, I had a trump card up my sleeve.

LOCAL KNOWLEDGE MAKES ALL THE DIFFERENCE

It was while I was developing the website for North Molton that I had mentioned to Mac of the North Molton History Society that I was mindful of writing a book about the gold found in the area. I had asked if there was someone local I could contact who had could provide me with any knowledge of the subject. Mac immediately, told me that there was one person in North Molton who could help me. His name was Norman

Govier and he was an expert on the subject. So it was when during the evening of 10th May 2012, as I was putting pen to paper, or rather writing my book in my software Kindle Writer that I had developed, I gave Norman a ring.

Norman answered the phone and I introduced myself. Before long we were lost in discussion about the mine and the reported gold found there. I got the impression he was as excited as I was about my call and he was most helpful in explaining some issues that remained outstanding for my book. As for Norman himself, he had quite a story to tell about his life and his interests in the mine.

AN INVITATION I COULD NOT REFUSE

As our discussion began to come to a close I asked Norman when it was that he would be visiting the site of the mine next. To my great delight he said that he was visiting the site tomorrow (Friday) delivering some sand to the mine. I did not know why he would be delivering sand to the mine, but in my excitement I did not ask him but requested instead if I could tag along. Norman said that would be OK! To say that I was elated is an understatement.

Although I was a total stranger ringing up out of the blue Norman knew that I was the person who built the North Molton website, and besides Mac of the History Society had stopped him a couple weeks before telling him that he might receive a phone call from me. So my call was not totally unexpected.

We agreed to meet in the village square about 11.00am, the next day. That night was a restless one as the excitement of visiting the Bampfylde mine played on my mind. What was I going to see? Would I find any gold? Not likely I thought but this did not dampen my exhilaration at the thought of visiting the mine or my anticipation of the next day's adventure.

THE ADVENTURE BEGINS

Friday dawned and after getting my digital camera ready, my micro cassette recorder to take verbal notes, a flask of tea and some sandwiches, I set off on my adventure and what an adventure it would prove to be.

I arrived on time at the North Molton village square and parked my car. All was quiet as I looked around for Norman to come. I had expected him to meet me in his vehicle but instead, I met him a few minutes later on the corner near the village shop. He had walked from his home that was a little way down the road.

There was a chill in the air and although the weather forecast was to be sunny in the afternoon, black clouds threatened to shed their load. Norman was well prepared though, wearing a waterproof lined coat and wellington boots. I asked him about the sand and he said that he would deliver that another day. Getting into my car he directed me to Heasly Mill and within ten minutes we reached the hamlet. Norman directed me to park in the old school, now closed that was on our left.

Heasley Mill school car park. The school was closed long ago.

I parked the car and putting on my wellingtons and a yellow waterproof plastic cagoule over my fleece, we proceeded to walk to the end of the hamlet along what was the main street. Needless to say from the moment we had met, we had talked about the mine. It all sounded very exciting.

I MEET THE YOUNGEST SON OF SIR HUGH STUCLEY

As we walked up the main road of Heasly Mill, passing what was once the mill to the left that gave the hamlet its name, I could hear the rumble of a vehicle approaching. Ahead of us was a narrow road rising up a hill which was obscured by the woodland that covered the area. Suddenly, a very large blue tractor appeared on the road in front of us and as we moved to one side to let it pass it stopped. A happy smiling man seated in the cab leaned forward and greeted Norman. It was clear that they knew each other well, judging from the banter that passed between.

The man on the tractor was none other than Peter Stucley, the youngest son of Sir Hugh Stucley, 6th Baronet, and Lady Stucley, of Affeton Castle, near Crediton, Devon. It is Peter who adminsters the land upon which the Bampfylde Mine stood.

Norman introduced me and Peter reached down and we warmly shook hands while Norman told him about the book that I was writing. Peter still smiling told me that I was in good company as Norman was the only person he trusted with unrestricted access to the Bampfylde mine accept during the time of the pheasant shoots that took place from time to time for obvious reasons. It was then that I really appreciated the incredible piece of luck that had befallen me to been taken on a tour of the mine by the only man who could do so with the knowledge, authority and the respect of the owner of the mine. I was certainly in good hands.

How did the Stucley family become involved with the mines? Briefly, in the early part of the sixteenth century the descent and ownership of the manor of North Molton had become a matter of a long-standing dispute. To cut a long story short the dispute was eventually resolved when Richard Bampfylde of Poltimore proved that he was the sole legal heir of the estate. He was duly given possession of most of the land and properties of North Molton by the Crown. The male descendants of Richard Bampfylde have inherited the North Molton estate ever since. When copper was eventually discovered on the estate the mine was called Bampfylde mine, after the family name.

In 1831 a descendant of Richard Bampfylde namely George Warwick Bampfylde was made the first baron Poltimore (1786-1858) and it is he to whom I have referred to in the previous chapter as Lord Poltimore. One of his descendants was George Wentworth Warwick Bampfylde who became 4th Baron Poltimore (1882-1965). He had a daughter Sheila Bampfylde, (1913-1996) who married Sir Dennis Frederic Bankes Stucley (1907-1983) 5th Baron of Affeton, Devon. The couple had five children (John, Rosmary, Christine, Sarah and Hugh). John was stillborn so Hugh inherited the title and became the 6th Baron of Affeton upon his father's death.

Sir Hugh Stucley married Angela Caroline Toller, the daughter of Major Richard Charles Robertson Toller, in 1969 and she became Lady Stucley. The couple live at Affeton Castle and Hartland Abbey in Devon. They have four children, two sons and two daughters. Peter is the youngest son of the couple and George Dennis Bampfylde Stucley is the heir apparent to the Affeton estate. The daughters are Lucinda and Charlotte Stucley. Peter is therefore the great grandson of George Wentworth Warwick Bampfylde.

Peter Stucley has written an article in the book published by the North

Molton History Society under the heading "Bampfylde and Stucley. Family Memories". It makes interesting reading.

It must be emphasised that the land upon which the mine is situated is private property and nobody is allowed on the land without prior permission from Peter or the Stucley family - Norman being the exception. Peter can be contacted through the Elms Estate Office, Bishops Tawton, Barnstaple, Devon, EX32 0EJ in writing or via a contact form on the website (www.elmsestateoffice.com).

ACCESS TO BAMPFYLDE MINE

I must confess I was very impressed by the friendliness of Peter and his evident genuine respect for Norman. It later transpired that Norman's grandfather had worked on the estate for Peter's grandfather, and so their roots had been intertwined for generations.

When Peter drove away, Norman and I proceeded along the road in the direction from where Peter had arrived. It was only a short distance to the track that led to the mine workings from where we stood. Crossing a small bridge that traversed the river, there was on our left a closed wooden gate that opened up on a dirt track, or in this case a muddy track because of the recent rains. Norman opened the gate and led me through. One would never have imagined that through this gate on the land beyond lay the once busy Bampfylde mine. My heart was pounding in anticipation of what I was about to see that was hidden from the eyes of most people.

The gate entrance to the mines

THE CRUSHER HOUSE

We walked along the muddy track with woodland trees and undergrowth on either side. There were strange rusty looking puddles all the way, and Norman said that they were coloured that way because of the iron content of the soil. We continued to talk about the mine as we walked and it was not long before we approached a clearing where a ruined building stood. This was the ruins of the Crusher House, where the copper ore excavated from the mining shafts were taken and crushed into smaller manageable pieces ready for transportation to Barnstaple to be shipped to the smelting furnaces around the country.

Norman with the Crusher House in the background

It was now that I learned what the sand was for that Norman had intended to deliver. He and a close friend Roger Burton were consolidating what was left of the building, and the sand was to be used for mortar. I was gob-smacked! Norman is 65 and Roger is 80 and here I had found two mining enthusiasts who had taken upon themselves to save the last remaining building of the Bampfylde mine. Between them they were restoring it for posterity supported by the land owner Peter Stucley.

The Crusher House is listed as a Grade II building, registered as such on the 24 November 1988, English Heritage Building ID: 398775. It exists today because both Norman and Roger Burton have volunteered hours of their own time, working with materials that they themselves have financed. A picture taken by M Hesketh-Roberts in 1996 show what the crusher looked like then. It was just a pile of stones overgrown with plants, hardly recognisable as anything but a pile of rubble. But thanks to the two aged enthusiasts the building has been cleared from undergrowth and much of the stonework had been repaired. The building looks quite majestic.

Left: Crusher House as photographed in August 1996 by M Hesketh-Roberts.
Right: The Crusher House the author photographed May 2012 in the process of being restored by Peter Govier and Roger Burton

Alongside the crusher house on the left was located a thirty foot diameter wooden water wheel, with a three foot axis. Not only did it drive the crushing machine in the house but also it was linked to a winder mounted on pillars to the west for the engine shaft. Nothing remains of the water wheel today only the open-air chamber where it was installed gives one an appreciation what an incredible machine it must have been.

Norman in front of the open-air chamber where the 30 foot water wheel was installed.
I have used some poetic license and reconstructed what the wheel might have looked like

NORMAN SHOWS ME SOME GOLD

It was early afternoon when we decided to sit down and eat our sandwiches. The weather had improved and it had brightened up. Fortunately, the threat of rain had not materialised. As I ate my sandwiches I cast my eyes upon the surrounding landscape.

The crusher house was set on high ground and looking towards the river Mole to my right I could also see some spoil dumps. These were heaped alongside the fence that separated the mining area from the river. There was evidence that someone had removed some of the spoil and Norman said that this had been done quite recently. Farmers had probably used some of spoil as hardcore for various building projects. Although nature had reclaimed some land I was surprised how much had remained untouched considering that over a 150 years had passed since the mine had closed, but then nothing can really grow on sterile rocks of the gossan that littered the mining area.

Scenes near and around the crusher house

The mine area was truly a tranquil place with just the noise of the River mole to disturb the silence. Norman told me how pleasant it was here in the Summer and it was very spectacular when he visited the site during the hard winter a year before when during Christmas we had heavy snow that remained on the ground for weeks.

During this period of peaceful bliss and while we ate our food Norman pulled out some rocks from his backpack. These he had found while walking around the mine area on previous visits. One by one he showed them to me describing each one as this or that with tongue twisting names that I was completely unfamiliar with. Then with a smile on his face and a cheeky look, he pulled out a rock. This glittered with what appeared to be veins of gold.

Are the golden coloured veins on Norman's rock gold?

Wow! Was this gold I asked Norman? Norman was non-committal and he said that he could not say. Of course, by now I was a believer and I had been struck down by a mild case of that strange malady known as gold fever. So as far as I was concerned anything that glittered was gold! The word "fool" comes to mind now as I write this sentence, a word that will take on greater significance later in the book.

THE ENGINE SHAFT

After we had eaten our lunch, we next went behind the Crusher House to view what used to be the mine workings. It is from here that the picture shown in the South Molton had been taken, although the vantage point was probably higher than where we now stood. The barren landscape of the mine workings as depicted in the picture have long since gone and the area is covered with individual trees and miscellaneous shrubbery where nature has been able to find fertile soil in between the sterile rocks of the dumps.

One could see a number of posts standing up in the distance but these have nothing to do with the former mine workings. They were used with respect to pheasant shooting, although I do not know in what capacity. I could also see a lot of spoil (gossan) where they stood and that is because this was the location where the engine shaft was situated. Here nature had failed to cover the scars that man had made.

We walked to where the engine shaft was located. Only a shallow depression marks the spot as the shaft has now been completely filled in. The top of a single wooden beam driven in the ground is all that remains to testify that a mine shaft had been dug here.

A pictorial overview of the main mining area of the Bampfyld mine as it appears today

To the right of the engine shaft and rising up a slope was a spoil dump that extended for about 400 yards if not more. Full of reddish coloured rock of all shapes and sizes, here nature had failed to get a foothold. At the top of the slope was fencing beyond which was woodland, that bordered South Radworthy field. In the field, although at this time I did not visit them, were two shafts. One went by the name of Field shaft - more about this later.

WHERE THE BERDAN MACHINES HAD BEEN LOCATED

A little further on, Norman took me to an area where nature had succeeded in taming. One could see old walls barely discernible covered in moss and vegetation. This was the place where the two Berdan machines had been installed for assaying gold from the gossan. This was a dangerous area because Norman warned me that the ground was heavily polluted by the mercury that was used by the machines. He suggested that we did not linger any longer and move on.

The location where the Berdan machines were installed.
This area is heavily polluted by mercury that had been used by the machines

END OF FIRST VISIT

It had been a most interesting and inspiring day. Norman had been the perfect host who clearly enjoyed showing me the Bampfylde mining area and describing the history of the place. Unfortunately, it was nearly time to go. However, we returned to the dump that backed onto the river Mole which recently had some gossan removed. Norman had said earlier that this was a good place to look for minerals because new rocks were now exposed for the first time and we might be able to find something interesting. My thoughts were centred on gold of course and so I did not need telling twice. What transpired next I shall reserve until later in the book.

After about twenty minutes of digging we decided to make a move to go home. What an exciting day it had been, but we were not finished yet. We went back to where the Berdan machines had been installed, and continued to follow the muddy track to the end of the mining area. We now walked along a narrow section, the river on the left and woodland screened by a high wire fence on the right. We came to a closed tall wired gate and a wooden bridge across the river. Norman, opened the gate and we went across the bridge, closing the gate behind us. We entered a small field and crossed it to the stone road bridge that crossed the river. We made our way back along the road heading towards Heasley Mill.

Before long we came upon a clearing on the right, a natural lay-by, where once horse carts had waited to transport the worker to the mine shafts at the top of the steep escarpment. To the left was a track that took the workers to shafts 3 and 4. It was very steep as I was to discover on another visit.

On the western side of the river is a natural layby created no doubt when mining was done here. To the left is a track that leads to Shafts 3 and 4 and to the right is a vertical trench where a 50 foot water wheel was constructed. An air adit is also here.

Norman took me to the right, passing a great scar of a vertical trench, now almost obliterated by foliage of all kinds. It was here that the 50 foot diameter vertical western pumping wheel designed by Captain Moorsom began to be erected in 1853 with the intention of powering the pumps in Engine Shaft. This wheel is referred to in a letter to the ,i>Mining Journal in February 1855.

> .... *There is a new wheel, not yet completed .... fixed on the face of the hill on the west side of the mine and about 40 fathoms distant from the pumping shaft .... to be connected by a horizontal beam working on supports at about 40 feet above the level of the road. .... the water to drive this new wheel is to be brought from a distance of more than 1¼ mile by a new leat which has been cut for the purpose on the face of the hill ...."*

The letter was from a report made by Messrs Hand, Moffatt and Marshall, and it goes on to quote a further report by Messrs Henwood and Hensley which critised the location of the wheel, the poor construction of the pit and the lack of water supply for the leat. The latter assessment proved correct and the project was abandoned but not before much of the wheel had been constructed at great expense.

We walked to the very end of the lay-by. Norman pointed to an adit, a horizontal tunnel that had been used for providing air to the underground mine workings. It had long since been sealed and was overgrown with all kinds of shrubs and foliage. A couple of fridges had been dumped in the depression in front of it which made it quite an eye-sore.

An adit that had been used for providing air to the underground mine workings can still be seen

What a fantastic day it had been. Thanks to Norman I had learned considerably more about the mines that had I relied entirely on research papers alone. There can be no doubt that a field study of the Bampfylde mining area had paid dividends in more ways than one, as I shall explain later. We walked back to Heasley Mill to the old school where my car had been parked and returned to North Molton. Norman invited me to his home, a neat bungalow that he had built himself. I met his wife Norma who made me a cup of tea, while Norman showed me his huge and very impressive mineral collection. Their hospitality was second to none.

Norman has a large collection of over 1000 mineral samples that he has collected over the years from around the world, some of which are quite spectacular. One side of his study beneath the window which has a wonderful unhindered view of the rolling hills of the village of North Molton is a long bench underneath which are numerous drawers. These contain his much beloved mineral collection. Behind on the other wall is his extensive library of books and documents.

As I returned home my head was still whirling with all that information that I Norman had shown me. I still did not know if their was really gold at the Bampfylde mine or not but the tour of the mine complex and what

Norman had told me all helped to put things into perspective and everything began to into fit into place. My book began to take shape.

I think now at this juncture that it is important to tell you more about Norman, his life and those that he befriended who shared in his love for minerals and his interest in the mines of North Molton. This is what I shall describe in the next chapter. It makes fascinating reading how the lives of three men became intertwined and the wonderful friendship they shared.

Chapter 6
THREE MEN AND A MINE

NORMAN MEETS JOHN ROTTENBURY | KINDRED SPIRITS

Like most boys caves and holes in the ground hold a special fascination and for Norman Govier to live in the vicinity of Bampfylde mine was like a moth attracted to a bright hazardous light. His family had worked on the Poltimore estate for several generations, so the mines were familiar to him and would remain so until this day. He regularly provides guided tours of the mine to interested parties and he certainly knows his stuff.

Norman's parents had repeatedly warned Norman of the dangers of the mines and forbade him to visit them. The mine complex was derelict and overgrown, and some shafts were still open. One slip and that would be that! But Norman ignored his parents warning and often he and his friends would bicycle over to the mines after school at North Molton and play. It was their special place where they could wonder freely without hindrance and do what most boys do at that age, have adventures. Little or no fencing prevented access at this time so the area was just a great place to play and to enjoy.

The mines with depths of 100 to 300 feet were certainly dangerous but thankfully nobody slipped to a watery grave in the darkness below. So just like my grandson (and me secretly) who love throwing stones into the stream running in the woods behind our house, Norman and his friends enjoyed throwing down large rocks down the shafts to hear the long lasting echo of the splash when they hit the water deep below. Often Norman would return home with red stained hands and clothing to receive a lashing tongue from his parents. He could not fool them where he had been. Boys will be boys.

NORMAN MEETS JOHN ROTTENBURY

This fascination for the mines of North Molton and minerals in general would remain with Norman all his life. Numerous opportunities would come his way to pursue his interests and one of these occurred in his early twenties (c. 1968) when his father introduced him to John Rottenbury.

John Rottenbury was a local farmer who had an extensive knowledge of the local mines and mining. It was part of his mineral collection which I saw at the South Molton museum. When John was in his forties and with his interest and wealth of knowledge about the mines at North Molton he captured the attention of British Kynoch Metals. The company offered him a position to carry out feasibility studies to reopen old mines around the world and one of these was the Bampfylde mine. We shall learn more about the feasibility study for this mine in a later chapter.

While working for British Kynoch Metals the company sponsored John with a scholarship to attend Leeds University where he wrote a thesis for his doctorate Ph.D in 1974, called *Geology, Mineralogy and Mining History of the Metalliferous Mining Area of Exmoor*. Although often quoted by researchers, the thesis was never published. While he attended the university, his wife and son ran their farm in his absence.

Returning to our story, in 1968 Field Shaft was one of the few remaining shafts within the Bampfylde mine complex that had not been filled in, and John Rottenbury wanted to take advantage of the opportunity to explore it one more time before the inevitable happened. So he was intending to go down Field Shaft that was located in South Radworthy field to the east of the mine and wanted someone to take photographs of the event. Norman was volunteered by his father because at this time he had become a keen photographer.

We visit the location of Field Shaft high up and too the east of Bampfylde mine in South Radworthy field. Here we see Norman viewing the shaft now filled in which he and John Roddenbury decended and explored c 1968

At the location John Rottenbury had set up a somewhat precarious arrangement to lower himself and Norman down the shaft. Attached to a hand powered winch that was roped to some nearby trees was a cable of seemingly inadequate diameter that secured a large industrial metal bucket. Slowly and at one at a time John and Norman were lowered down the shaft to about 145 feet. They could not go any lower because the water level was at 300 feet. Here John explored a horizontal adit while Norman took some photos.

John Rottenbury 145 feet down Field Shaft (Picture Credit: Norman Grovier)

It was the first of a number of enjoyable visits to mines around the Exmoor area that Norman and John Rottenbury shared over the years until their friendship came to an untimely end when John sadly passed away at a relatively young age.

KINDRED SPIRITS

Norman's interest in minerals and mines in particular has ever remained foremost in his mind. So it was that years later when his son was at university and with his wife's approval he joined the Exmoor Mines Research Group (EMRG). The group was formed as a society in 1992 following a symposium on "Mining on Exmoor" hosted by "The Exmoor National Park Authority". Since then the group had met informally as a band of friends who share an interest in mining and industrial history.

Norman joining the EMRG led to several industrial archaeology digs on the Brendon hills with visits and talks on other West Country mines. It was through the group that he met who was to be his great friend Roger Burton.

Roger Burton is the author of two local books *The Heritage of Exmoor* and *Simonsbath: The inside story of an Exmoor Village*, as well as numerous articles for the EMRG magazine. Roger has spent many, many hours researching a wealth of information on local history, mining and mining families. At the time of the publication of this book I have not had the opportunity to meet Roger. But it is thanks to him through Norman that I have been able to obtain facts for this book, including a report from a survey that was carried out in 1989 that has been vital in the proving whether gold is to be found at Bampfylde mine.

Left: Friends Roger Burton and Norman Govier. Right: Roger Burton and EMRG colleague
Taking part in Exmoor Mines Research Group activities.
(Picture Credits. Roger Burton)

It is through Roger Burton that Norman joined the Combe Martin Silver Mines Preservation Society. The society was formed in 2001 to act as a support group to the Combe Martin Miners who have been active from 1989 at their Mine tenement site engaged in the restoration of mine buildings, the excavation of the engine house, the chimney flues, Williams Pumping Shaft, and Harris's Mine Shaft, all having been filled upon the discontinuation of mining in 1880's.

The group meets twice a week working, both above ground and 90 feet below the surface, where they are opening up and strengthening the old tunnels. Norman can be found at the mine most Sundays volunteering his services.

Norman is now retired and the wheel has turned full circle. He regularly visits the Bampfylde mine providing tours to interested parties and with the support of the Exmoor Mines Research Group together with Roger Burton he is helping to consolidate the old crusher house at the Bampfylde mine. Roger Burton now 80 (2012) still works at the top of the scaffolding as long as Norman can get the material up to him. They have done a remarkable job, I am sure you will agree.

The Crusher House at the Bampfylde mine which Norman Govier and Richard Burton are consolidating

Chapter 7
THE BAMPFYLDE MINE REVISITED

SHAFT No 3 | SHAFT No 4 | I SEARCH FOR GOLD

When I visited the Bampfylde mine with Norman for the first time, I was impressed by the size of the mining area, but I had only seen half of it. There were still extensive mine workings on top of the steep escarpment on the western side of the Mole river that I had not as yet seen. If you recall we had briefly taken a look at a natural lay-by on this side of the river and saw the great overgrown vertical trench where the 50 foot water wheel had been built, and a horizontal access adit. I had no idea what to expect when a couple of weeks later we returned to the site to do a field survey of the shafts in this area.

SHAFT No 3

Unlike the main mine workings on the east side of the Mole river that are in the valley, Shafts 3 and 4 and other workings can be found on the top of a heavily wooded escarpment. To get to them there is a wide muddy track that winds and zigzags up the slope with a gradient that must be greater than 45 degrees. It is quite a climb but upon reaching the top the view of the mine area below and across the river is breathtaking and certainly worth the climb. However, we were not here to admire the view but we were on a mission to take a look at what was left of the mining operations on this side of the river.

The first thing we saw when we reached the top of the track were the huge quantities of spoil that litter the place. Somewhere amongst the spoil that was piled so high was Shaft No 3, but finding it was difficult. There was a shallow depression where Norman thought the shaft might have been, but if this was the shaft there is no trace of it. It was filled in long ago.

SHAFT No 4

Moving on, we scrambled amongst the piles of dark-pink-red rock that was everywhere. From these we slowly made our way through thick undergrowth until we eventually came to Shaft No 4, not that I could easily identify that this is what it was until Norman pointed out various mine ruins lost to nature. The shaft itself was hidden in thick undergrowth and only a shallow depression in the ground testified to its former existence. It was now filled up with rubbish that had evidently been deposited there by farmers. There was no point in taking a photo as nobody looking at it would believe that it was anything but a rubbish tip.

The spoil dumps of Shafts 3 and 4. Little else remains of the shafts or the mine workings

If Norman had not been my guide I would not have appreciated what was on the ground. A loose railway track lay amid the grass, upon which ore carrying trucks had once travelled. Broken walls hardly visible because of moss and creeping plants covering them littered the scene. One wall was at least 8 foot tall standing in tangled undergrowth and nearby four iron posts stood erect in a square, their original purpose lost to history. Who would have thought that on this derelict site that hundreds of workers had once trod the very ground upon which I was now standing.

We lingered for a while but did not stay long. Even so we were so high up that we could see the countryside for miles. Norman pointed to various spots where other mining activities had taken place, and in the distance we could see a helicopter moving along electricity pylons and cables doing a visual inspection.

Traces of mining paraphernalia at the location of Shaft No 4 on the western side of the Mole river

I SEARCH FOR GOLD

We returned to the huge piles of rock that was near Shaft No 3. I wanted to see if I could find any gold here and for about half an hour I had the pleasure of searching through the rock but at this time my research had not been completed and I did not know what form the gold would be found. Needless to say nothing glittered to suggest that gold was here, and I came away empty handed. Even so, the visit had certainly been worthwhile, if only to enable me to appreciate the big picture of how the Bampfylde mine was laid out and how huge the area was that it covered. Here the rock differed from that on the eastern side of the river. It was heavier, harder to break and there appeared to be very little malachite or quartz to be found.

Chapter 8
THE TRUTH IS REVEALED

FELTRIM MINING SURVEY (1989) | BRITISH GEOLOGICAL SURVEY (1994) | BGS - ASSESSMENT | BGS - CONCLUSIONS AND RECOMMENDATIONS

Is there gold at Bampfylde mine? That has to be the million dollar question. What have we found out? There have been reports of gold being found at the mines of North Molton for centuries, with Bampfylde mine proving to be source for most if not all the gold findings reported in the North Molton mining area.

I can confirm that there was several rocks containing gold on display at the South Molton museum. I have seen and handled them and I was reliably informed that they had come from the Bampfylde mine. They had been donated to the museum by John Rottenbury who, according to Norman, had no doubts that gold could be found at the mine.

We also know that a scam took place at the Britannia mine, while at the same time we had identified that no such swindle occurred at the Poltimore Copper and Gold Mining Company at the Bampfylde mine. In fact this company itself was subject to fraud by an equipment supplier that had supplied mercury extraction machinery to the site which cost the company dearly and forced it to close prematurely. This resulted in misleading reports that this company too had been part of the same scam instigated by the Britannia Gold company, when in fact the company was merely the victim of fraudulent activities by the equipment supplier.

All this leads to limited tangible and circumstantial evidence as well as hearsay to show that gold had been found at the Bampfylde mine. This is not enough to state categorically that there is gold at the mine and probably the reason nobody has really made an effort to learn about the mines until I came along to write this book. However, we are in luck! In recent years a number of significant surveys have been carried out and these are described below. Do they support the evidence hitherto presented in this book? Read and be amazed what they say.

FELTRIM MINING SURVEY (1989)

What you are about to read is an unpublished report of a mining survey that took place early in 1989 at the Bampflyde mine by the Feltrim Mining company. I am grateful to Roger Burton for providing me with the documents upon which this particular investigation is based and his permission to reproduce and quote from them here.

In 1989 the company in question undertook what is known as a grab sample survey, removing 20 samples of what was classed as mineralised rock from spoil that lay on the surface of the mine area. Even today in 2012 there is a considerable amount of spoil remaining although a quantity has been removed for use, according to Norman, for farm building projects in the area.

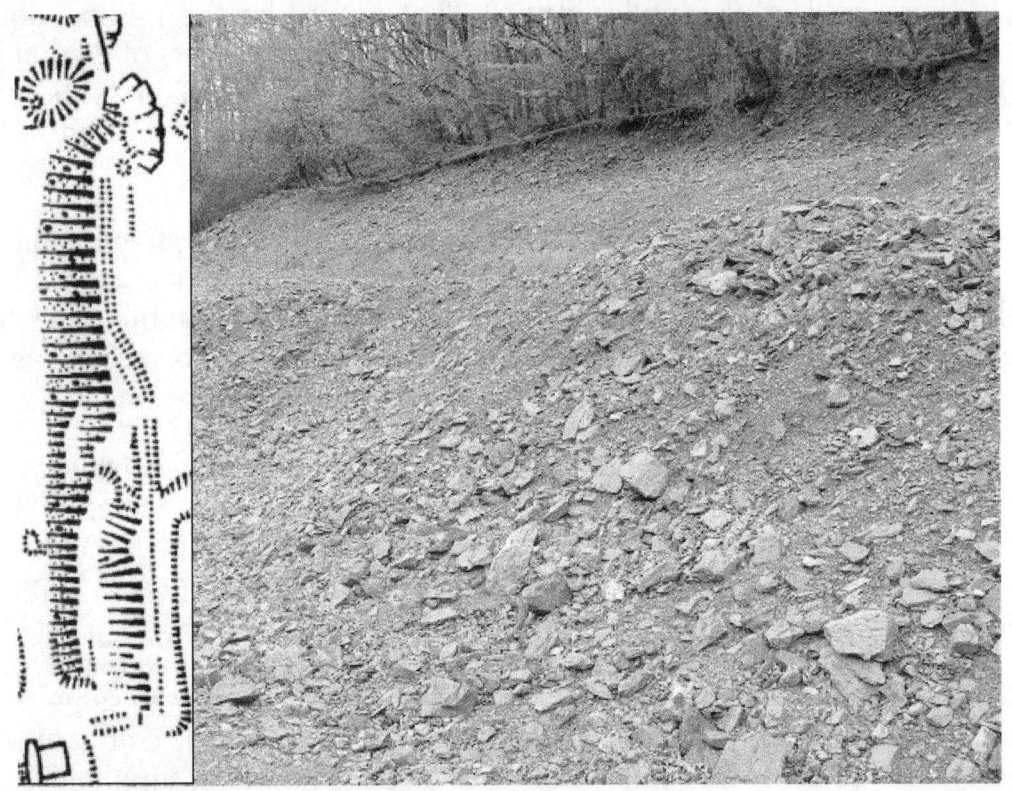

Large quantities of spoil exists even in 2012 as can be seen from this photo the author took. It is found on the right-side from where the main shaft existed.

The samples were then sent to OMAC Laboratories Ltd, in Galway Ireland for analysis and a certificate of analysis was issued on the 18th August 1989. The results were amazing! 90% of the samples (18 out of 20) contained gold with sample 15 = 3555 ppb and sample 16 = 1120 ppb.

OMAC Laboratories Ltd
Athenry Road, Loughrea, Co Galway, Ir
Telephone: 091-41741, 41457
Telex: 50985 OMAC EI
Eirmail: EIM 351
Fax: 091-42146

CERTIFICATE OF ANALYSIS   18th August 1989

TO: Feltrim Mining
ATTN: N. Haughey
CODE: NM/QRS

NM1 - NM 20: BAMPFYLD MINES: DUMP STUFF - MINERALIZED.

BATCH NO.   6F91
NO. SAMPLES 50   Rock

LAB NO.	SAMPLE NO.	Cuppm	Pbppm	Znppm	Asppm	Auppb	RepAu	RepAu	New Portion Auppb
1	NM 1	X	X	X	X	-20			
2	2	X	X	X	X	645	254	383	527
3	3	X	X	X	X	487		363	714
4	4	X	X	X	X	140			
5	5	X	X	X	X	357			
6	6	X	X	X	X	423			
7	7	X	X	X	X	125			
8	8	X	X	X	X	39			
9	9	X	X	X	X	-20			
10	10	X	X	X	X	21			
11	11	X	X	X	X	95			
12	12	X	X	X	X	32			
13	13	X	X	X	X	402			
14	14	X	X	X	X	67			
15	15	X	X	X	X	44			
16	16	X	X	X	X	3555	3305		
17	17	X	X	X	X	1120			
18	18	X	X	X	X	410			
19	19	X	X	X	X	930		380	668
20	20	X	X	X	X	840	337	173	201

Letter to Roger Burton with results of OMAC samples. Only the first 20 samples are shown as they relate to Bampfylde mine only.

What does these sample readings mean in layman's terms? When we talk about ppb (parts per billion), the formula is 1,000 parts per billion =1 gram per tonne of gold. Hence, sample 15 had the equivalent of 3.555 grams of gold per 1 tonne of ore. This is quite substantial. In financial terms and based upon the purity of the gold of the sample, according to gold prices today (21st May 2012 - DI KJ Noble Metals, Watford, Herts) a tonne of ore from where sample 15 was taken would equate to between £ 37.16 (9 carat 37.5% purity) to £ 93.14 (99.9% purity).

carat*	weight	amount paid
9 ct/ carat = 37.5 % purity	1 gram	£ 10.460
10 ct/ carat = 41.6 % purity	1 gram	£ 11.653
14 ct/ carat = 58.5 % purity	1 gram	£ 16.570
18 ct/ carat = 75.0 % purity	1 gram	£ 21.371
21.6 ct/ carat = 90.0 % purity	1 gram	£ 25.735
22 ct/ carat = 91.6 % purity	1 gram	£ 26.200
24 ct/ carat = 99.9 % purity	1 gram	£ 30.710

One tonne of ore taken from where sample 15 was removed, would be worth up to £ 93.14 (3.555 grams @ £30.71 today.

The question now arises, based upon the survey results would mining the spoil from the mine be worth doing? Here is what the *Encyclopedia of New Zealand* says about the matter in their article *Gold and Gold Mining*.

> "For alluvial deposits near the surface a payable grade can be considerably less than 1 gram per tonne. For hard-rock sources mined in opencast methods, such as at Macraes Flat, around 1.6 grams per tonne is just profitable at 2005 prices."

Since 2005 gold prices have rocketed. This would suggest that it would be profitable to mine for gold at the Bampfylde mine. Kenneth J. Gerbino Company would seem to agree, when talking about what would be profitable and what would not. He says:

> *"With a near surface potential open pit gold deposit, 2 grams per tonne (a gram is .03215 of an ounce) would be excellent. 1 gram would be fair as long as you don't have to remove too much waste rock to get at the ore."*

Clearly, if the gold was simply taken from the existing spoil, of which there are huge quantities at the Bampflyde mine, then mining would be potentially profitable. Now we come to a mystery. If this was what was reported to Feltrim Mining, why did the company not pursue this seemingly lucrative opportunity? Why indeed.

This very question was raised by John Hamilton the man who undertook to take the samples in 1989. In a letter to Roger Burton dated 30th June 2000. He said:

"At one stage, not too many years ago, a rapid grab sample was carried out the Poltimore mine and of the 20 samples taken from the dumps - all mineralized material, only two gave background values in gold. All the others were certainly anomolous and I cannot imagine why the mining company did not ask me to go back and follow up on the initial work."

The author having done some digging, excuse my pun, has found out the reason why Feltrim Mining did not pursue this great opportunity. At the time when the samples were taken at the Bampfylde mine, Feltrim Mining was in significant financial difficulties. It became necessary for the company to make critical changes to its financial infrastructure. The information that follows comes from the Independent of Ireland, 21st May 2012 under the heading *Minmet investors set to roast top brass at EGM*.

Feltrim Mining was floated on the Irish Stock Exchange on the 1st of April of 1988. From their accounts approximately one million pounds was raised on the flotation of the company bringing in much needed financial support. The shares were originally offered at 40 pence and then the share price doubled to 80 pence on the first day of trading but fell back steadily thereafter.

In the first year of trading the Board had approved expenditures of some £600,000 on various exploration ventures, and it was probably at this time that samples were taken at the Bampfylde mine. Unfortunately, for the company their ventures came to nothing, which did not help their financial position. In 1989 the company incurred a loss of £838,804. Things only got worse because in the following year there was a deficit on the profit and loss account in excess of 2 million. Disaster loomed for the company.

It does not require rocket science to understand why Feltrim failed to invest in the Bampfylde mine. The company was in no financial position to do so. It is as simple as that.

That there is gold at Bampfylde mine cannot now be doubted but only ten samples had been take at the mine by Feltrim and the investigations of one John Rottenbury that also confirmed this earlier may not satisfy some critics. They could argue that the people concerned just got lucky. What is needed is a comprehensive survey by an industry recognised body to substantiate the findings. Fortunately, the author discovered such a body who had done just that - the British Geological Survey.

BRITISH GEOLOGICAL SURVEY (1994)

In 1994 the British Geological Survey undertook to survey seven abandoned iron and iron with copper mines in the North Molton area, with a view of finding out if there was the possibility of economic amounts of gold in the waste tips of those mines. One of these was the largest mine workings of the area namely Bampflyde mine. Nearby Britannia, Crowbarn and New Florence mines were also surveyed. The survey report is listed as Release No 15: **An appraisal of the gold potential of mine dumps in the North Molton area, North Devon** and written by D G Cameron and D J Bland (1994).

The British Geological Survey (BGS) had observed that John Rottenbury had reported that discrete gold had been found in the soft matrix of secondary materials such as gossan or malachite. On the basis if this BGS undertook to extract gold through physical means rather than make an assessment based solely on chemical analysis. Consequently, grab samples of approximately 4kg was collected for analysis.

According to BGS of all the mines surveyed, ONLY the Bampfyde mine spoil, both on the east side and the west sides of the river Mole contained gold. Twelve samples had been take over the entire area and the location of each sample was also recorded. From the samples the BGS could estimate how much gold could be extracted from the individual spoil tips. They estimated that 872 Troy ounces could be extracted from tips on the west side of the river, and 330 Troy ounces on the east bank. This would equate, according to the report, to an average of 701 Troy ounces over the entire site. This is quite phenomenal as we shall discuss later in Chapter 10.

ASSESSMENT, CONCLUSIONS AND RECOMMENDATIONS

I have paid for and obtained permission by the British Geological Survey to reproduce the final assessment, conclusions and recommendations of their survey for this book. Coming from such an esteemed professional body, I believe that their survey provides an honest, informative and unbiased assessment of the Bampfylde mine site.

I first start with the assessment:

ASSESSMENT

Gold was only recorded in material from Bampfylde. No trace of gold was found in the sample from Britannia Mine and none of the samples from the other sites were auriferous.

At Bampfylde, significant quantities of free gold are present on the dumps. This could be extracted using bulk concentrating techniques, e.g. using a Knelson Concentrator, without much pre-processing of the dump material except for screening to a suitable size. There may be more 'bound gold' in the >2 mm size fraction of the dump material which could be liberated by additional crushing.

At the time of the 1974 survey, an estimate of 85,000 cubic yards was obtained for the remaining dump material. It is not known how much material has been removed since that date but local information suggests that most of the material remains on site.

An attempt to assess the amount of gold in this material is presented below (Table 5). The following assumptions are made in order to make this assessment:
a. The samples are representative of the dump and the dump represents run-of-mine material.
b. No gold was lost during panning.
c. The 25% reduction in the weight of gold is balanced by the underestimation of grain numbers.
d. The analytical result for the chemical gold content of the >2 mm size material and weight of grains from the <2 mm size fraction can be amalgamated.
e. The silver content of the gold is low.
f. A figure of 70% recovery by physical means is used.
g. The amount of material in the dumps is still 85,000 cubic yards.
h. The average density of this material is assumed to be 3 g/cc.

Table 5: Estimate of quantity of gold in Bampfylde dumps

West side Sample No.	*Average g/t	Amount of gold in dumps at this concentration Troy oz at 70% recovery	East side Sample No.	*Average g/t	Amount of gold in dumps at this concentration Troy oz at 70% recovery
131	0.173429	760	134	0.234513	1027
132	0.64775	2838	135	0.122965	539
133	0.00356	16	140	0.015	66
138	0.2925	1281	141	0.039156	172
139	0.0525	230	142	0.010679	47
144	0.025	110	143	0.029238	128
Average W Bank	0.199123	872	Average E Bank	0.075259	330
* calculated from combination of concentration from <2 mm and >2 mm fractions in the proportion 0.25:0.75			Average all sites	0.0160056	701

British Geological Survey © NERC. All rights reserved. CP12/053

What now follows is the British Geological Survey conclusions and recommendations.

"Gold is present in appreciable amounts in the tips at Bampfylde Mine...." - BGS

CONCLUSIONS AND RECOMMENDATIONS

1. Gold is present in appreciable amounts in the tips at Bampfylde Mine, but not in the other tips sampled.

2. Assuming 70% recovery and 85,000 cubic yards of material, c. 700 Troy oz of gold might be extracted from the tips at Bampfylde Mine. However, as the variation between these small samples is large, the amount of gold that could be extracted might vary between less than 100 and over 2000 Troy oz.

3. Further work at Bampfylde would be necessary to prove this resource, including assessing a methodology for commercial extraction.

4. Further reconnaissance sampling needs to be carried out on the East Buckland, Wheal Charles and Uppcott mine sites. Britannia Mine site should also be sampled in more detail.

British Geological Survey © NERC. All rights reserved. CP12/053

The full report can be obtained from the BGS at the following website address:

Release No. 15: An appraisal of the gold potential of mine dumps in the North Molton area, North Devon

Chapter 9
THE ICING ON THE CAKE

WHAT IS THIS? | ANOTHER DISCOVERY | TENTATIVE GOLD TESTING | THE THIRD VISIT TO BAMPFYLDE MINE | THE MOMENT OF TRUTH

When I began this book I had no idea where it would lead me. It had begun with an obscure sentence in a book published by the North Molton History Society. Had I not developed their website it is doubtful if I would have heard about the mines because even my local museum had no information on them. It ended with indisputable proof that gold existed within the spoil of the Bampfylde mine and more than that, in substantial quantities.

It has certainly been a roller coaster of a ride, but my story does not end here. What I have deliberately left for last was what transpired on the day of my tour of the mining area with Norman and thereafter. It was truly the icing on the cake.

WHAT IS THIS?

My first day at the Bampfylde mine had gone very well, and I had learned a great deal from Norman about the history and geographical layout of the mine workings. I have to admit that there is nothing better than gaining an overall perspective of the mine than from actually being there on the ground in order to put flesh to the bones to the drawings and notes that I had accumulated in the course of my studies.

We were coming an end of the tour by which time we had now descended a slope west of the ruined Crusher House. There was a lot of spoil piled up here, but what was of particular interest was a depression in a dump of gossan that lay against the fencing that separated the mine areas from the river Mole. It had recently been dug out and Norman said that this had been done quite recently.

This was good news. It meant that possibly we might find something that had lain hidden beneath the spoil and which had now been exposed to the surface as a result of the dig. The opportunity to be able to search for gold bearing rocks was too good to pass by and so I scanned the ground in the hope that I might see a glint of gold.

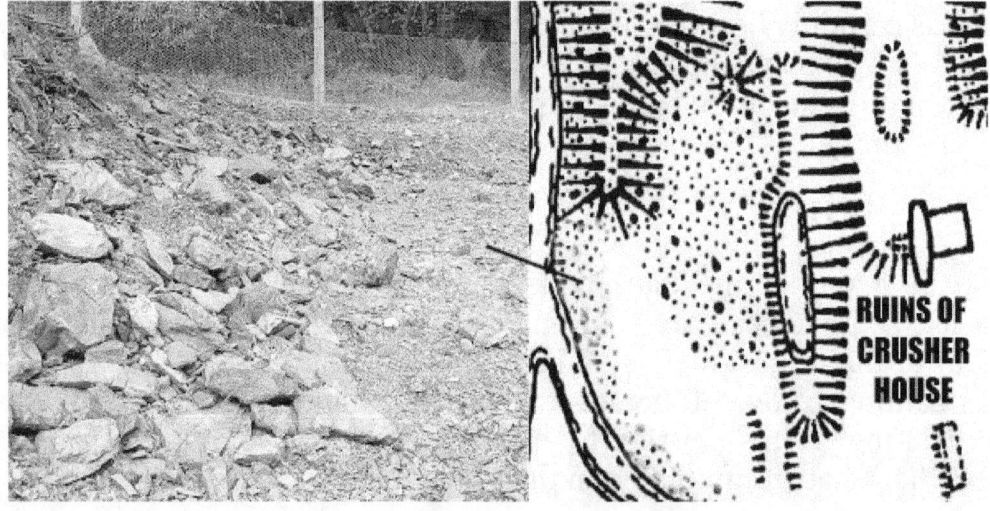

We both started digging around the area, I with an elongated rock to dig with that I had picked up from the ground and Norman with his digging hammer. For a while, although there were some interesting rocks to be found including those that contained malachite, a bright green mineral associated with copper, I did not find any rock that contained any gold.

After about 20 minutes of searching it seemed that we had been wasting our time if it was gold that we were looking for. Then my heart missed a beat. Under a large rock that I had moved was a smaller one about 1 inch long and 3/4 inch wide and on this I could see a glint of gold at each end. I could not believe my eyes. I picked it up and gasped with excitement. What I saw certainly looked just like gold.

I turned to Norman and said to him that I thought that I had found some gold. I reached out my hand to show him, only to find that he was doing the same. Smiling, he had in his hand a smaller rock about half-inch square. At the end was quite a large piece of the same looking gold that I had found. He had just found some too.

I am sure you can imagine how I felt. I was elated - but was it gold or was it fool's gold? It could have been fool's gold but I have not seen any reports of such in any documents that I had researched. One would have thought that even at the height of the gold fraud at the Britannia mine, where one might have thought that fool's gold might have been used to convince people that there was gold in the mine, nothing was said. The odds were in my favour.

Another reason I believed that what I was looking at gold was because one of the rocks in the South Molton Museum that had been identified and tested and said to show gold looked very much like the same type of rock as the one that I had found.

Rock Comparision: Left (South Molton Museum) Right (My Sample)

The day at the mine had come to an end and not only had I learned a great deal more about the mine from Norman thanks to his tour of the place, but here I was with two pieces of rock that appeared to contain gold. Needless to say, when I drove home that day I was over the moon. Little did I know then that at home I would make another discovery.

ANOTHER DISCOVERY

Besides the two rocks that appeared to contain gold I had also taken home a mixture of rocks that just looked interesting. About a week or so later I had the idea of breaking up those rocks to look inside. When we were on the Bampfylde site this is something that I had not done so it was just something that I needed to do to satisfy my curiosity.

Using a steel hammer I selected a few rocks that resembled the one that I had found with gold on edges. Well knock me down with a feather! In one of the rocks I found more gold-like material embedded within. It was almost too good to be true. The findings of the British Geological Survey which said that *"Gold is present in appreciable amounts in the tips at Bampfylde Mine"* seems to have been born out by my own very tiny grab survey.

TENTATIVE GOLD TESTING

I now had four rocks that appeared to contain gold, two which I had originally found with gold insitu and two halves of another one that I had broken up. The problem was that although circumstantial evidence suggested that I was looking at gold, it was still circumstantial and it was possible that what I was really looking at was fool's gold. While gold is a precious metal, fool's gold (also known as pyrite or iron pyrite), is not a metal. Rather, it is a made of a crystaline structure made from iron and sulphur that has a metallic luster with a gold colour very similar to gold. I needed some means of determining that I had found gold and not fool's gold.

Incidentally, the expression "fool's gold" doesn't refer to just any old fool, but originally referred to the Queen of England. During colonial times, some intrepid British explorers wanted to establish a new colony on the coast of Labrador, but were denied the funding because they had not found any gold there. Not to be deterred, they collected some fine specimens of pyrite and sent them back as proof of a gold discovery, and the ruse worked! The Queen looked at the samples (but not too closely) and immediately approved the building of the colony. It was a hundred years before the trick was discovered.

I carried out some more research on the Internet and found that the most common testing method for gold was to use nitric acid. Nitric acid is a highly corrosive and toxic strong mineral acid which reacts with most metals and it is this characteristic that has made it a common agent to be used in acid tests, such as the one I was needed for testing gold. It just so happens that gold and platinum group metals do not react with nitric acid. Perfect!

I managed to purchase a testing kit from Amazon which consisted of 5 bottles of acid of different strengths. Two were for testing Silver and Platinum, the other three were for different carats (Kt) of gold. These were 9-14Kt, 18Kt and 22Kt solutions.

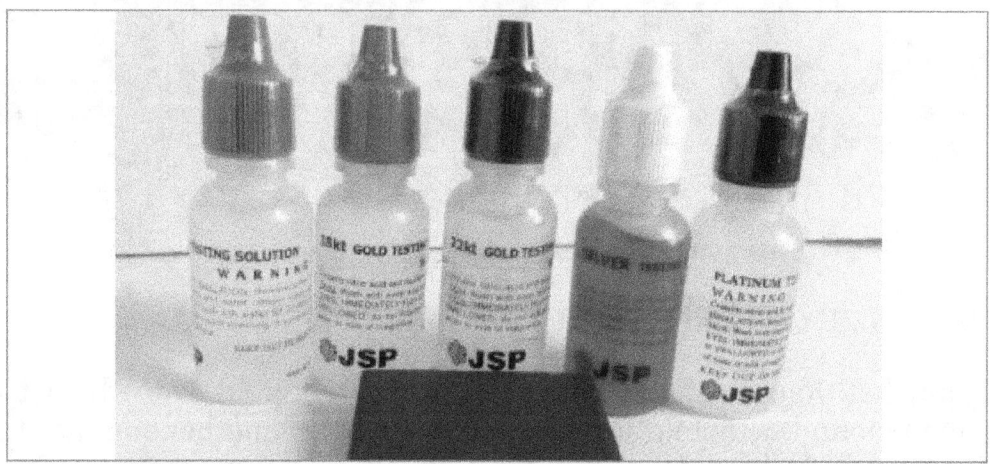

Gold Comparison Table

Karats	Compostion	Gold %
24k	24 Parts Gold	99.9%
22k	22 Parts Gold, 2 Parts Other*	91.6%
18k	18 Parts Gold, 6 Parts Other*	75%
14k	14 Parts Gold, 10 Parts Other*	58.3%
10k	10 Parts Gold, 14 Parts Other*	41.7%

* Metal alloy, such as copper, nickel or silver

The instructions were not at all clear. They were aimed at testing jewellery, spoons etc and were not designed for testing gold in rocks. Anyway, I visited Norman in North Molton as he had a sample of fool's gold, sourced elsewhere. In our naivety we assumed that fool's gold would react differently to that of gold. We applied the 22Kt solution to the fool's gold and nothing happened. There was a slight change I noticed namely that the fool's gold appeared to gain a more silvery luster. Norman thought that this was because the acid had dissolved some rock and therefore the crystals of pyrite were more exposed than before. In any case the test on the fool's gold was not what I had expected. I had expected it to dissolve or change colour or something, but t do anything was a big surprise. We next tried nitric acid on my sample of "gold" and it too did not change much either although it did appear to me that the gold took on a more golden colour than before. What was going on?

What we did not know was that Nitric Acid has no effect on fool's gold as I was to learn later on a website:

> "Place the substance in nitric acid. Nitric acid does not dissolve or tarnish raw gold. Fool's gold, however, also is not affected by nitric acid, but other identification methods will help you determine whether or not the specimen is fool's gold."

Furthermore, our testing methodology had been incorrect. We were looking for changes in the mineral on the surface of the rock under test and what we should have done was to wipe the acid from its surface with a white paper kitchen towel and see what colour it was stained with. More about this later.

To say that I was very disappointed is an understatement. The tests had proved to be inconclusive and besides my specimens were tiny. I need to obtain more specimens for testing. So I arranged to go back to the site with Norman again specifically to spend more time searching through the spoil for more gold bearing rock. By now I had a good idea what kind of rock to look for.

THE THIRD VISIT TO BAMPFYLDE MINE

I had been to the Bampfylde mine twice so far. The first was the original tour with Norman to obtain an overview of the mine workings. This was also the tour when I had found my first samples of gold, or at least what I thought was gold. The second visit was to explore the west bank of the river and visit Number 3 and 4 shafts. While we were at this location we did spend some time looking for gold in the massive spoil dumps that littered the area but the rocks were unlike the ones in which I had found the gold specimens. The rocks were heavy and reddish in colour, some containing quartz and malachite.

I needed to find more samples of gold so at the end of May 2012 Norman and I returned to Bampfylde mine for the third time. As before we parked my car at the school at Heasley Mill and walked once again along the footpath that took us to the spot where I had found the gold near the ruined Crusher House.

On the way I took the opportunity to take a look at Field shaft, the one that Norman had gone down with John Roddenbury, so many years ago. It was all filled in now, with barely a trace to say that it had ever been a mining shaft. There were some spoil round about but nothing of interest caught my eye.

Moving on we reached the spot where I had found my original sample of gold. I was eager to put my theory to the test. I searched for the same kind of rocks that contained my original sample of gold and within minutes I had struck gold! For another 20 minutes or so, I broke open similar rocks and before long I had quite a collection of possible gold samples.

Norman was surprised how easily I had found the 'gold' especially, when I broke open a particularly large rock and found the largest specimen yet. It was about 3 inches wide and 2 inches long. I gave half of the rock to Norman for him to add to his mineral collection, and I continued to dig for smaller samples. Later at home I broke apart the remainder of the big rock into two pieces and within was an incredible amount of gold-like material.

THE MOMENT OF TRUTH

I had by now quite a large number of samples of rocks that appeared to contain gold. Even so I still did not know if it was gold or not. As the acid tests had been inconclusive what I needed to do next was to have my samples analysed by an electronic gold tester. These are very expensive, but I was aware that some jewellery shops sometimes have them. So the following day I went into Barnstaple to see if I could get the samples tested.

One shop had an acid testing kit just like mine and they carried out a test. This was then that I had discovered then that my previous testing methodology had been incorrect. Upon putting a drop of nitric acid on the surface of the rock where the gold was situated, the jeweller then wiped the acid from the surface with a paper towel and compared the resultant colour to that of a colour chart which denoted what colour would appear with different grades of gold or not. To my relief she expressed the opinion that what I possessed was 14 karat gold or better. That was fantastic but I needed a different kind of test just to make sure.

After visiting a number of jewellers I found one that had an electronic testing kit for testing for gold. This was a Tri-Elecronics THE G-XL-18 described by the manufacturer as *"The best, most advanced, and most reliable electronic gold tester in the world. It eliminates the need for acid tests and it reliably measures gold karattage to the nearest karat in the 6 to 18 karat range. For gold items above 18 karats the unit reads 18 karats or more."*

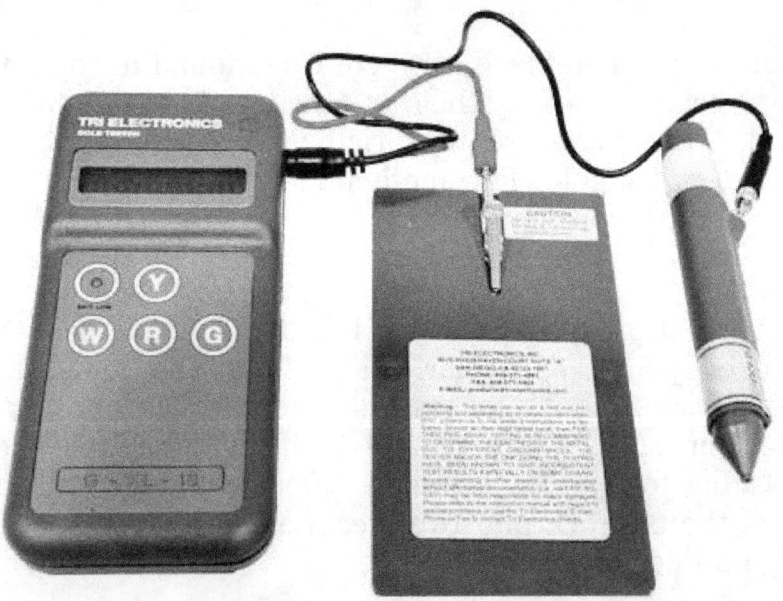

The Tri-ElecronicsTHE G-XL-18 electronic tester used to test rock samples.
It is described by the manufacturer as "The Best, Most Advanced, and Most
Reliable Electronic Gold Tester In The World"

The lady in the shop was most helpful. She could not test the gold that was embedded in the rock as it had to be place inside a crocodile clip. However, I did have three thin shards (from rocks that I had broken) and she was able to use these. The moment of truth had arrived. She placed one of the shards in the crocodile clip at the top of the unit and using a probe she measured the gold. It was 14 Kt gold. She then tested the other two shards and the result was also 14kt gold.

GOLD!
There's Gold in them Hills

Acid Test: 14 Kt gold
—Electronic Tests—
Sample 1: 14 Kt gold
Sample 2: 14 Kt gold
Sample 3: 14 Kt gold

Why 14 Kt gold? The gold is probably pure gold but because it was tested on copper bearing rock, it was seen by the electronic tester (and acid tests) as being an alloy and hence reported as such, a mixture of copper and gold.

I wondered why it was 14 karat gold. I would hazard a guess that the gold was probably pure but because it was tested within copper bearing rock, it would have been seen by the testing systems as an alloy, a mixture of copper and gold. Regardless of the purity I had found gold! Yippee!

Chapter 10
THE STING IN THE TALE

SOMETHING IS NOT QUITE RIGHT | THE PHONE CALL | FURTHER TESTS | THE ACID TEST | THE GLITTER TEST | THE HARDNESS TEST | I RETURN TO SOUTH MOLTON MUSEUM

For centuries there had been rumours of gold having been found in the mines of North Molton but it is only now as a result of this book that it can be said with certainty that those rumours were true. In fact I can also say with confidence that substantial quantities of gold can be found in the spoils of the Bampfylde mine even today. I can say this because this has been confirmed by two surveys that has taken place within the last 30 years. Furthermore, one of these was carried out by a recognised and reputable geological institution the *British Geological Survey* and their conclusion was that, *"At Bampfylde, significant quantities of free gold are present on the dumps"*.

It should be noted that free gold had only found at the Bampfylde mine, the site I have visited with Norman. The BGS did not find any gold at the Britannia mine the scene of the scam that I described in an earlier chapter that took place in 1853 despite claims to the contrary. Likewise, no gold was found at the Crowbarn and Florence mines that lay further south of the Bampfylde mine.

Besides proving that substantial quantities of gold could be found at the Bampfylde mine, I had also demonstrated that the claims by the **The Poltimore Copper and Gold Mining Company** that they had excavated gold at the Bampfylde mine has been substantiated. The company unfortunately was caught up by the fallout from the Britannia mine gold fraud and became tarnished with the same brush.

We now come to the question of the 'gold' that I had in my possession and which I had found at the Bampfylde mine. Was this further proof that gold was to be found on the gossan dumps at the Bampfylde mine workings? It would seem so as it had been independently tested by two jewellers who had identified that I had samples of rocks that contained 14 karat gold. I phoned up Norman and told him the news. He was still not convinced.

SOMETHING IS NOT QUITE RIGHT

For a few days I languished in a kind of ecstasy as the collection of gold glinted at me from the box I had placed the rocks. Not only had I proved that gold existed on the dumps of spoil at the Bampfylde mine but also I had physical proof to show to anybody who asked. I had every reason to be smug.

Even so there was a nagging feeling that something was not quite right. It all seemed to good to be true that I had in my possession for all intents and purposes quite a lot of gold and it had been easy to find. In fact I was already thinking about notifying Peter Stucley of my finds which I felt obliged to do because the gold had been found on his land. However, I decided to wait a little longer until the book was published and then I could approach him to discuss it with him when all the facts were out in the open. It is just as well that I delayed because of what was to transpire next.

The nagging feeling I had about my box of gold persisted. You see, I had read what Peter Claughton had said in his paper, *Gold and North Molton*. He had said:

> *"What had attracted attention was the metal exposed in the oxidised zone of the lodes near surface. Here copper had been leached away leaving the iron gossan in which were small amounts of free gold, some large enough to be visible to the eye."*

Another writer, Chris Ralph, of whom I will speak about later made a similar comment.

> *"Most people think of gold nuggets and such as the source of gold, but the truth is that very little of the new gold produced comes from nuggets - nearly all newly mined gold comes from ores mined from the natural hard rocks that contain gold in tiny, even microscopic particles.*

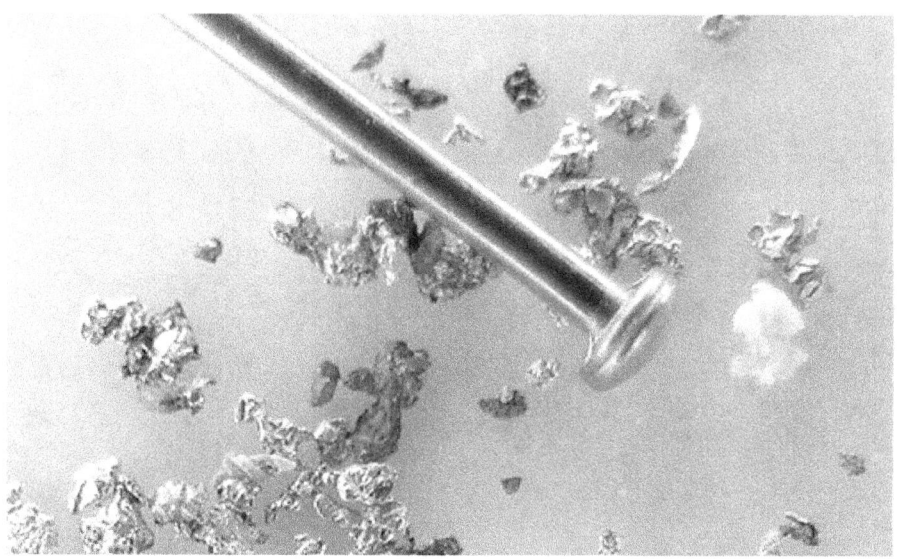
Tiny grains of gold barely visible to the naked eye is what should be found in the spoil dumps of the Bampfylde mine. This is not what I had found.

Do you see what I mean? Observers were saying that gold today comes from tiny, even microscopic particles and this was said to be true of the Bampfylde mine too where grains of gold were described as being barely visible to the naked eye and if they were they would only seen as tiny grains. My box of gold simply just did not fit into this picture. Each sample had quite a bit of gold on its surface.

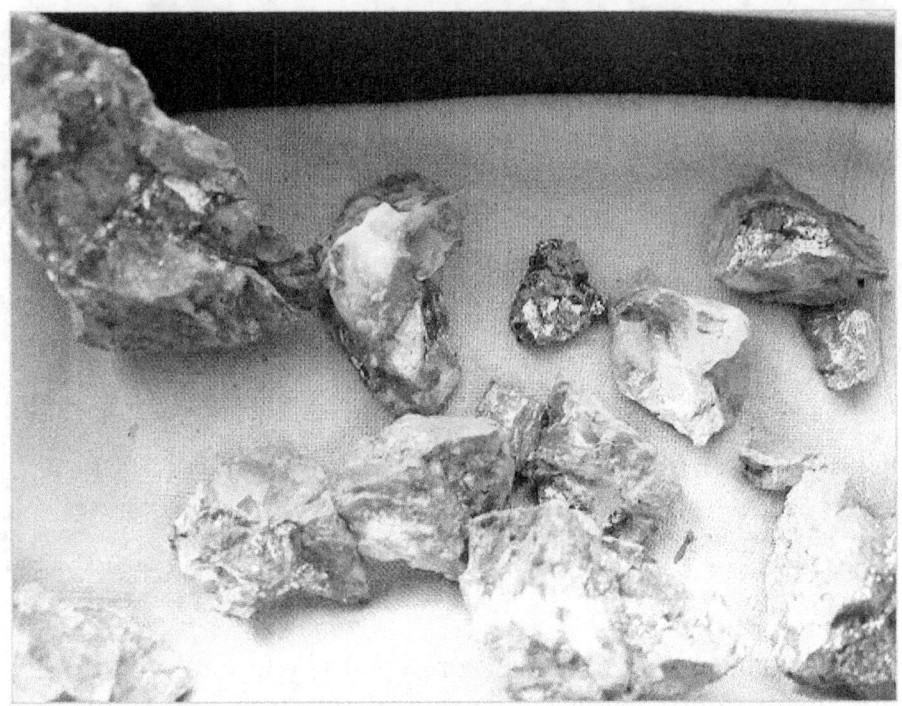

I had found lots of gold but according to reports what I should have found was small grains, and most of these were said to be invisible to the naked eye

Finding the gold had been easy - too easy. Once I had recognised the type of rock that I should be looking out for it was very easy to go back to the site, search for the same kind of rocks and break them open to find gold within. My success rate had been very high. But I was not finding one or two grains of gold. I was finding a whole mass of gold grains accumulated together in various sizes. Something was not quite right. Yet! I had the results of acid tests and electronic testing by independent sources confirming that what I had in my possession were rocks containing 14 karat gold. What more proof did I need? I convinced myself that the low purity to be due to perhaps the detection of another metal, probably copper, and hence the tests showed up my gold as an alloy and not pure gold.

THE PHONE CALL

Two days after my third visit to the mine I received a phone call from Norman. He had not been convinced that what I had in my possession was gold and he had carried out some tests of his own. He had placed the piece of rock I had given him in a makeshift furnace. After a while, the gold in the rock had turned black. This was not what one would not expect to find happening if it was gold. Gold is a metal and it would simply have melted. It was quite evident that I did not have any gold. Norman was certain that what I had was fool's gold.

To say that I was stunned is an understatement. This was not what I had expected to hear. Even though I was beginning to have my doubts too, I was still starry eyed and suffering from a "gold fever". I was sure that what I had found was real gold. Was it not true that independent tests had proved it? Norman had to be wrong!

Such were the thoughts that crossed my mind but one could not ignore the fact that Norman has been involved in mineralogy for most of his life and he was truly an expert on the Bampfylde mine, its history and geological structure. As a newcomer I was hardly in a position to argue with his findings. What I needed to do was to carry out more tests so that there could be no room for doubt. So it was with mixed feelings I thanked Norman for his assessment and I said that I would carry out further tests myself. The question was what other tests could I make?

FURTHER TESTS

I searched the Internet to see if I could find out any other tests I could make that would prove one way or another whether I had gold in my possession. It took quite a while to find but having tried various search strings, I discovered a website that provided me with the answers I needed. Called *The New 49'ers - True Life Gold Prospecting Adventure* it was a mine of information. I wished I had found it earlier, it would have saved me a lot of money in buying a nitrate acid kit which was to prove completely useless.

The website is owned by Dave McCracken and he has a great deal to say about identifying gold from fool's gold.

> *"It is not unusual for a beginner to wonder about the difference between gold and the other materials found inside of a streambed or lode deposit. Sometimes a beginner will puzzle over shiny rocks; and quite often, iron pyrites (fool's gold) or mica are mistaken for the real thing. In fact, this is so much the case that there is a story of an entire shipload of iron pyrites having been shipped over to England from America during the 1500's, the yellow stuff having been mistaken for gold. So you can understand where it gets the term "fool's gold."*

This was the shipment from Labrador that Queen Elizabeth 1st had been fooled into thinking was gold.

It was interesting to read on the website that gold taken directly from a lode is crystalline in structure and is usually referred to as "rough gold" because of the coarseness of its surface. I could certainly say that the gold that I had found was course if not crystalline. This was good news. Dave McCracken also said that *"In its native state out of a lode, gold is almost never 100-percent pure, but has a percentage of other metals along with it"*. This would certainly account for the tests that had been carried out on my gold having been rated 14 karat. Again things were looking good for my gold portfolio.

Dave now described three tests to ascertain whether we are looking at gold or fool's gold. *"If you are just starting and have not yet had the opportunity to see much gold in its natural form, there are three easy tests which will validate your discoveries one way or the other."*

THE ACID TEST

> **Acid Test:** *Nitric acid will not affect gold (other than to clean it); whereas, it will dissolve many of the other metals found within a streambed. Nitric acid can be purchased from some drug stores or prescription counters, and can sometimes be found where gold mining equipment is sold. If you question whether your specimen is some metal other than gold, you could try immersing it in a solution of nitric acid. If your specimen is gold, it will remain rather unaffected. If it is most any other kind of metal, it will dissolve in the acid."*

I had already tried the acid test and sure enough the gold was unaffected, so it would appear to be gold. It is now that Dave throws a spanner in the works.

> *"Nitric acid will not affect iron pyrites or mica (fool's gold)"*.

The acid tests that I had implemented had not effected my gold samples. But like gold, if it was fools gold, it would not be effected either. The acid tests therefore were useless in determining whether I had gold or not. I hoped that the next test may be more conclusive. The glitter test as it was called certainly sounded interesting.

THE GLITTER TEST

> ***Glitter Test:*** *Gold does not glitter. It shines. Sometimes it is bright; sometimes it is dull; but very seldom does it glitter. The thing about fool's gold (pyrites or mica), is that because of its crystalline structure, it tends to mostly be of glittery appearance. Take the sample and turn it in your hand in the sunlight. If it is gold, the metal will continue to shine regularly as the specimen is turned. A piece of fool's gold will usually glitter as the different sides of its crystal-like structure reflect light differently.*

I took my samples and rotated them in bright sunlight. One rock part of which I had given to Norman, the one that he had put in the furnace, did indeed glitter - as did most of the other rocks I had. But one piece did not and furthermore the colour was different. Norman's sample rock had an almost silvery glint to it that glittered when seen from different angles. Not so the rock of which I had. It appeared to have a uniform shine at any angle. But as I turned it I could see that there were small areas of glittering material that differed from the gold that I was observing. The colour looked the same as Norman's rock and thus these areas evidently contained fool's gold, according to the glitter test.

It would appear that most of the rock samples that I possessed were in fact covered in fool's gold. However, it would appear that at least one rock contained both real gold and fool's gold on its surface, at least that was my interpretation. Was that possible? Fool's gold with real gold on the same rock. I did more searching on the Internet and soon found the answer. According to Chris Ralph who writes on small scale mining and prospecting for the ICMJ Mining Journal said in his recent article *The Source of Gold - Its Ores and Minerals*, he said:

> *"In various gold ores, the native gold commonly occurs as tiny particles contained within sulfide minerals such as pyrite. Iron pyrite is an exceedingly common mineral associated with gold, but it also serves as a reducing agent."*

This statement opened up possibilities that I had not as yet thought about. What Chris Ralph was saying in effect was that it was quite common to find tiny particles of gold **within** pyrite and that gold and fool's gold often coexisted. In other words, there could be free gold grains contained within my samples of fool's gold that was so tiny that I could not see them. It should be noted that all the gold samples that I had all were composed of individual grains compressed together so it was quite possible that some of those grains were real gold. This was quite a revelation. My samples may be fool's gold, but there may well be microscopic grains of gold within and this is what the tests that I had done had revealed.

The *Mining Journal* of 1847 also says that fool's gold can also contain genuine gold within it. The journal writes:

> *"Mr. Mitchell, who has a very extensive practice in assaying for the mining companies, says, expressly, that very many of the samples of iron pyrites he has seen contain both gold and silver, but the working of them is neglected."*

More recently, Michael Fleet writing in the *American Mineralogist* (Volume 82, pages 182-193, 1997) under the heading, "Gold-bearing arsenian pyrite and marcasite and arsenopyrite from Carlin Trend gold deposits and laboratory synthesis" says:

> *"In sediment-hosted gold deposits and some mesothermal lode-gold deposits, a substantial fraction of Au is present as "invisible gold" within grains of iron sulfide and sulfarsenide minerals (principally pyrite and arsenopyrite). Previous studies have established that invisible gold is spatially associated with local As enrichment in pyrite grains (Wells and Mullens 1973; Fleet et al. 1989; Cook and Chryssoulis 1990; Bakken et al. 1991; Fleet et al. 1993; Arehart et al. 1993; Mumin et al. 1994; Michel et al. 1994).*

Although I had by now more or less accepted that what I possessed was fool's gold (with perhaps the odd grain or two of gold inside), there was one final test one could make to prove whether I had found gold or fool's gold. According to Dave McCracken this was the hardness test.

THE HARDNESS TEST

> ***Hardness Test:*** *Gold is a soft metal, like lead, and will dent or bend when a small amount of force is applied to it. Pyrites, mica and shiny rocks are generally hard and brittle. Just a little amount of pounding will shatter them. Gold almost never*

shatters!

Everyone knows that gold is a soft metal, but my problem was that the rocks that I possessed were covered by tiny grains of crystalline material the colour of gold. Gold as it has been noted, when taken directly from a lode, can be crystalline in structure and this is known as "rough gold" because of the coarseness of its surface. But then so is fool's gold. The difference is that gold grains are irregular in shape while fool's gold, being crystalline, are not.

Using my 60x magnifying glass I was unable to determine the shape of the individual grains on my collection of rocks so it was now as case of scraping some of the "gold" from the rock onto some white paper and see what happened if I crushed them.

From several of my rock/gold samples I scraped off some gold coloured grains. I next proceeded to crush the scrapings and found to my surprise that they shattered very easily. Having crushed the 'gold' to a powder, the glint of gold had disappeared and what remained was a pile of black dust.

I searched the Internet for further information. Talking about flakes of gold, a contributor to the goldrefining forum wrote:

"if they just break up and 'flake' into black dust it might be that you have pyrite (or fools gold) , or mica , whatever you will."

Rock Currier on the Mindat.org forum, the largest mineral database and mineralogy reference website on the internet says:

Pyrite also has the bad habit of producing "black dust" when it rubs together. The black dust is just finely divided particles of pyrite that look black."

Finally, Hal Burton writing in his book, *The Real Book of Treasure Hunting*, says it all.

"Gold flattens; pyrite crushes to a black powder.

As I looked at the small pile of black powder on the white paper upon which I had crushed what I had thought was a piece of gold, I had to face the fact that the rocks that I had brought back from the Bampfylde mine were fool's gold. The pressure test had confirmed this conclusively. Norman had been right all along. But then perhaps not...

I RETURN TO SOUTH MOLTON MUSEUM

It was now evident that most if not all my rock samples were composed in part of fool's gold and not real gold. But having said this it was also possible that within or alongside the pyrite covered surface of my rock samples tiny grains of real gold may also exist and it was these that had been identified by the independent tests aforementioned. OK! I admit it. I may be clutching at straws but unless I spend a lot of money sending my rocks for proper analysis in a mineralogical lab which I cannot afford, I guess this is something I will never know for sure.

While there can be no doubt, based upon the recent surveys undertaken, that substantial amounts of gold exists on the dumps of the Bampfylde mine, it is not of the kind that you can pick up like nuggets. By all accounts gold only exists as tiny grains, mostly invisible to the naked eye. That being the case I was faced with a conundrum. The authenticity of the rocks that I saw at the South Molton museum and which formed part of the Rottenbury collection was now cast in doubt. I needed to go back to the museum and take a second look, and this time I would take Norman with me.

It was on the 12th June 2012 a little more than a month after my first visit that I returned to the South Molton Museum. Outside I met Norman and together we entered the doors of the Guild Hall, the entrance to the museum. Phil Tonkins was waiting for us. Phil recognised Norman immediately as he had attended a number of lectures that Norman had carried out in the past. Soon they were lost in chat remembering the good old days.

South Molton Museum: Phil Tonkins and Norman Govier deep in conversation

Inevitably we turned our attention to the rocks in the Rottenbury Collection. Phil opened up the display as before and both Norman and I had the opportunity to examine the rocks. The lighting was not very good so we were given permission to take the individual rocks outside to examine them in natural light. We first took the two smaller rocks that resembled my own collection, and it was clear to both of us that they were fool's gold too. Phil said that it was possible that these rocks had originated elsewhere and had got mixed up with the Rottenbury collection, although they were displayed with a note saying "Gold Bearing Gossan with Malachite - Bampfylde T1442".

My main interest however, was primarily focused on the large rock, the one that is shown on the cover of this book. I had mistaken the large layer of shiny mineral on the surface of the rock as gold. But this both Norman and I agreed was fool's gold. However, when I took a closer look with my 60x magnifying glass I could see what appeared to be irregular shaped grains of gold embedded inside. In view of the reputation of John Rottenbury I doubt that he would have been fooled by fool's gold and that what I was really seeing, barely discernible to the naked eye, were real grains of gold.

In Rottenbury's rock there appears to be embedded tiny irregular shaped grains of gold

As far as this particular rock was concerned I believed I could see genuine grains of gold. Norman was not convinced, but even he conceded that it was unlikely that his friend John Rottenbury had been in error.

It was time for me to go, and so I thanked Phil for his hospitality and left, leaving both he and Norman talking about old times. As for me I had got my answer, the gold that I had collected at the Bampfylde mining area was not real gold but fool's gold - but probably containing grains of 14 karat gold too. As I returned home with this revelation, I must confess I was not overly disappointed. For me, it did not matter whether I possessed real samples of gold from the Bampfylde mine or not although that would certainly have been the icing on the cake. What did matter was that I had set out to discover whether the tales of gold being found in the North Molton mines had been true and I am sure you will agree, I had achieved that objective.

Chapter 11
THERE'S GOLD IN THEM HILLS

WHERE IS THE GOLD TO BE FOUND? | THE ESTIMATED VALUE OF THE GOLD | A SMALL SCALE COMMERCIAL ENTEPRISE? | THE KNELSON BATCH CONCENTRATOR | THE ALL IMPORTANT QUESTION

It has to be said that there is no gold to be found at the Britannia, Crowbarn or New Florence mines. Any references saying that there is are unfounded and based upon falsehoods and misapplied information. However, with respect to the Bampfylde mine this is a completely different matter entirely. If truth be known considerable quantities of gold exists within the huge piles spoil dumps that litter the mine area on both sides of the river Mole. Of this the reader can be of no doubt upon reading this book.

If when I speak of gold you may be thinking of nuggets forget it. There are none to be found. The reported nugget found by William Flexman in 1810 originated elsewhere and not in the mines of North Molton. It is a fact that gold found these days are rarely found in the form of nuggets. The gold exists as free tiny, even microscopic particles, trapped within the rock. The grains of gold are so tiny that if you were to brake open any auriferous rock the gold would be hardly discernible to the naked eye. This is what all accounts of gold found at the Bampfylde mine describe.

I should have carried out my research more thoroughly because had I done so I would not have made 'fool' of myself when, overcome by a case of gold fever, I hoarded and coveted fool's gold. Fool's gold is an apt name for the gold looking pyrite you must agree. I should have read what Chris Ralph had to say in his article *The Source of Gold - Its Ores and Minerals*.

> *"Most people think of gold nuggets and such as the source of gold, but the truth is that very little of the new gold produced comes from nuggets - nearly all newly mined gold comes from ores mined from the natural hard rocks that contain gold in tiny, even microscopic particles*

Although I had been fooled by fool's gold I can tell you one thing. It was just as exciting to search for pyrite as it was if it was the real thing. It had been great fun breaking open rocks that I could easily identify as probably containing the glint of gold, and to be proven right. Now I can say that there is a lot of fool's gold to be found in the spoil heaps of the Bampfylde mine especially if you know where to look and what kind of rock one would expect to find it in.

Perhaps I shall go down in history as the first person to have ever reported finding fool's gold at the Bampfylde mine because I had not seen any reports by anyone else in all the many documents that I search during my investigations. However, it is not fool's gold that this book is all about. It is about real gold and there is a considerable amount to be found on the gossan heaps at the Bampfyle mining area.

WHERE IS THE GOLD TO BE FOUND?

The biggest mistake the **Poltimore Copper and Gold Mining Company** made was not to appreciate the patchy nature of deposition of free gold scattered on the site. They had a very important clue when an early sample assay revealed where gold in quantity could be found. *The Mining Journal of 1853* records:

> *"In the meantime an assay of gossan from the western side of the river at the Poltimore gave a return of 11 ounces per ton, encouraging both companies to continue with plans to raise gossan in large quantities."*

This early assay report pointed the way to where good quantities of gold could be found - the western side of the Mole. However, because of the chaos and hubbub of the mining operations, and the management of the company not keeping careful records, this fact was overlooked.

The directors of the company as amateurs were not very organised and although it was known that the sample had originated on the west bank, they did not know exactly where. The auriferous rock was transported from the top of the hill on the western side where the shafts were located by a system of overhead buckets to the main workings on the east side below - similar to the chair lifts in Switzerland. These buckets were simply filled from spoil in a haphazard way. The company should have logged where the auriferous rock was excavated and recorded the results when the assays had been done. This they did not do. It would be a costly mistake and their ultimate undoing.

When the first major shipment of 100 tons of rock was sent to Liverpool for assaying, it is evident that the rock shipped had come from wherever it was easiest to get at the mine. Most therefore came from the eastern side mixed with material from the western side. So when the report came from the Liverpool assay that said, *"The initial results in Liverpool were varied, one of 6 dwt per ton and another considerably higher at 10 ounces per ton."* we should not have been surprised. As a consequence, not knowing whence the best sample had originated, the company blundered about and repeated the same haphazard process again and again.

If the reader may recall in an earlier chapter of this book I said that I could now with reasonable certainty point to where the best returns of gold can be found. Would you like to know where the best quantities of gold can be found? You have already seen one clue as described above - the early sample of the Poltimore company. This and a careful analysis of the documents in my possession confirm that the greatest concentration of gold is to be found on - **THE WEST SIDE OF THE RIVER MOLE**.

THE WEST SIDE IS WHERE MOST GOLD CAN BE FOUND

It is evident that the greatest concentration of gold is to be found in the dumps on the west side of the river Mole. The largest sample of the British Geological Survey in 1994 (BGS 132) was found in the tip that was south-west of a western shaft (Number 4) and opposite the field boundary. The rock contained small amounts of grey-green shale, quartz, hematite, siderite, malachite. There was about 100 small to medium gold grains that were panned from the crushed rock.

Another large concentration sample was obtained from the same general area (BGS 131). It too contained small amounts of green shale, hematite and malachite as well as tarnished pyrite. A large sample (BGS 138) found at the nose of the roadside tip above the track crossing from north to south as well as within that tip (BGS 139) also contained a lot of gold. The rock consisted of purple, red, green sandstones, with quartz veins with hematite and malachite.

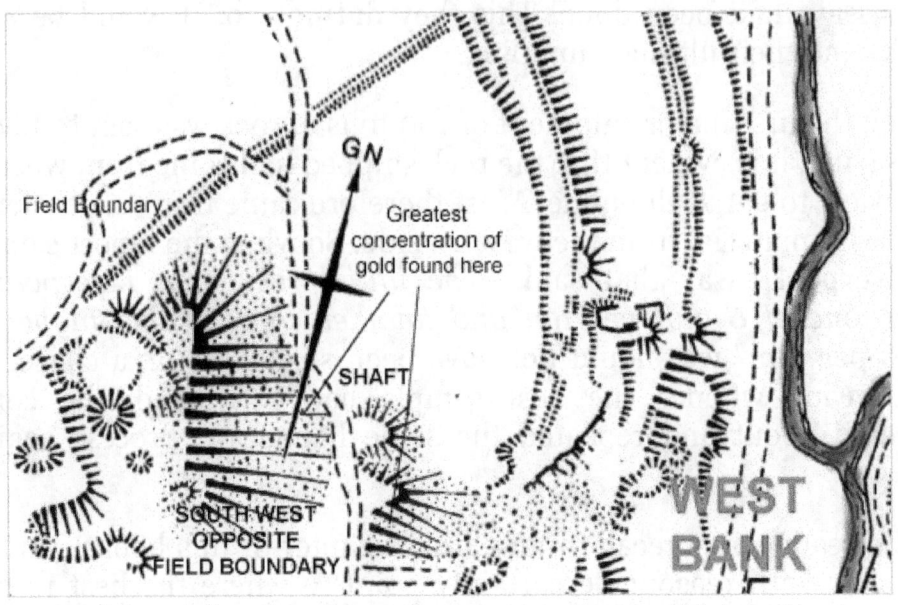

My interpretation where the greatest concentration of gold is to be found on the west bank of the Mole. Opposite field boundary, south-west of shaft. Also point of track crossing north to south

However, the book *Gold: Its Geological Occurrence and Geographical Distribution* by James Maclaren published in 1910 by the *Mining Journal* gives us conflicting information.

> *"The auiferous gossan-load is a firable ironstone, highly mineralise and containing copper. It is brown on the western side of the Mole and reddish on the the eastern bank. The latter portion of the vein is reputed to be twice as valuable as the former, assays giving 17 dwts. and 8 dwts. gold respectively."*

For the reader's reference twenty dwts of gold equals one Troy ounce so we are talking about quite a lot of gold here.

> The chief auriferous gossan-lode is from 4 to 10 feet wide, and dips to the north. There is considerable evidence of this mine having been worked, probably for copper, in very remote times. The auriferous gossan is a friable ironstone, highly mineralised, and containing copper. It is brown on the western side of the Mole and reddish on the eastern bank. The latter portion of the vein is reputed to be twice as valuable as the former, assays giving 17 dwts. and 8 dwts. gold respectively.
>
> From the book "Gold: Its Geological Occurrence and Geographical Distribution" by James Malcolm Maclaren published in 1910 by the Mining Journal.

It is true that James Maclaren contradicts my analysis by saying that the greatest concentrations of gold exists on the east bank of the river and not the west. However, one needs to keep in mind that he was reporting on data recorded 40 years after the event and it makes a lot of sense to assume that most of the auiferous gossan-load had been processed on the eastern side of the river as it was close to where the processing machinery was located. Also, he used the term "reputed" so writing in 1910 James Maclaren obviously had no written documents to back up what he said.

Furthermore, most of the gossan dumps have gone on the east side of the bank of the Mole and all that remains are the large dumps that lie alongside the edges of the South Radworthy field. Hence, it should be no surprise therefore to find that in 1994 the *British Geological Survey* found that the best concentrations of gold on the eastern side come from these dumps. And, because so much of the auiferous rock had been removed elsewhere on the eastern bank of the river this would explain why BGS was able to record better results on the western side where much of the dumps still remained in-situ.

It is worth noting that the colours of the gossan dumps on either side of the river by James Maclaren would appear to be an accurate assessment judging from the photos I took when visiting the dumps on both sides of the river.

Spoil dump on the western bank of the Mole river. More brownish

Spoil dump on the eastern bank of the Mole river. More reddish

THERE IS GOLD ON THE EAST SIDE TOO

Even today there is an area on the eastern bank of the Mole where there is a reasonable concentration of gold to be found. This is the dumps on the edge of the South Radworthy field, which is east of the Crusher house. According to the *British Geological Survey* the most concentrated of gold were found in samples (BGS 134 and 135) and the rock consisted of heavily stained grey green slate iron ore, abundant hematite, tarnished pyrite and some malachite.

PYRITE (Fool's Gold)   **HEMATITE**   **MALACHITE**

Minerals found at the Bampfylde mining site

Hematite is a very common mineral, iron oxide, Fe_2O_3, occurring in steel-gray to black crystals and in red earthy masses: the principal ore of iron. As such it is also magnetic, and it is often used in jewellery. Malachite is a very popular mineral with its intense green color and beautiful banded masses. Polished, banded Malachite has been carved into ornaments and worn as jewelry for thousands of years, and in some ancient civilizations it was thought to be a protection from evil if worn as jewelry.

It was in the eastern side of the Mole river where I had found the fool's gold (pyrite), near the old crusher house that Norman and Roger were renovating. There was some malachite to be found and hematite, but I found more on the western side.

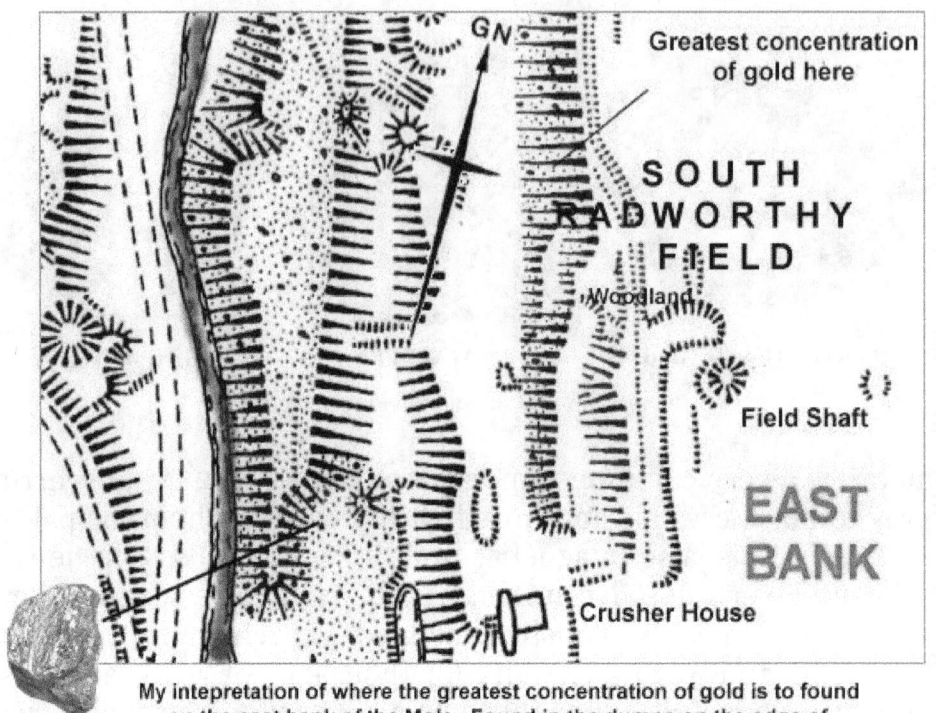

My intepretation of where the greatest concentration of gold is to found on the east bank of the Mole. Found in the dumps on the edge of the South Radsworthy field, by fence and wood

Fool's Gold Here

One should keep in mind that **ALL** the 16 samples taken by the *British Geological Survey* contained gold and only 4Kg had been collected for sampling. These samples came from both sides of the river with the western bank showing twice as much gold than the eastern bank. Likewise, the 20 samples taken for analysis for the *Feltrim Mining* company, with the exception of two that also contained gold of varying in quantity with samples 16 to 20 containing substantial amounts. This particular group of samples were probably obtained from the same general area and since these were the last samples selected one can reasonably assume that they came from the western side of the Mole river, as it would have been natural to start to take samples within the main mining area on the east bank first.

It is clear that if a company was to return to the Bampfylde mine with the intention of mining for gold the place to get the best returns would be on the western side of the Mole river. However, the ore dump running along the edge of the South Radworthy field on the east bank of the river should not be ignored. The question we need to ask is, would it be economically viable to do extract fold from the spoil heaps today?

THE ESTIMATED VALUE OF THE GOLD

Large quantities of gold exists on the spoil dumps of the Bampfylde mine of that there is no doubt, but how much is it worth in financial terms? Assuming that the 1994 British Geological Survey total average estimate of 700 troy ounces of gold is accurate then the present value of gold can be roughly calculated.

According to the gold calculating website "Goldprice Org" (2012) one standard ounce of gold is worth £1033.27. However, BGS used troy ounces in their figures of calculation and 700 troy ounces equals 768 standard ounces. Hence, we can therefore calculate that the amount of gold lying on the surface of the Bampfylde mine in spoil heaps could be worth no less than £791,821 - over three-quarters of a million pounds. That is a sum not be sneezed at.

The figure of 700 troy ounces of gold to be found at the Bampfylde mine by the BGS is only a conservative estimate. BGS also said in their report:

> *"as the variation between the small samples is large, the amount of gold that could be extracted might vary between less than 100 and over 2000 Troy oz."*

Those figures would therefore provide a total estimate of £100,000 to £2.5 million. Phew! Even the worst case scenero is not that bad.

Today, we have a significant advantage of the Poltimore company that worked the mines in 1854. We now know where the greatest concentration of gold can be found and as a result I would estimate that perhaps 1000 plus Troy ounces of gold could be produced, a sum exceeding £1 million. It is little wonder then that in the summary of their report, BGS came to the conclusion that there was a possibility of economic amounts of gold to be found in the waste tips of the mine. In fact on their website when introducing their report they commented further. They stated that their findings suggested that there may be sufficient gold present to support a small scale commercial extraction enterprise.

A SMALL SCALE COMMERCIAL ENTEPRISE?

How easy would it be today to process auiferous gossan to retrieve the gold from the Bampfylde mine? More to the point, how much would it cost for equipment and materials? Actually not very much considering the potential of **over** £1 million returns on the investment. I estimate between £80,000 to £100,000 based upon the following information.

Times have changed and so has the technology used to process gold. In the past enormous manpower was required to man the expensive bulky and heavy wooden/metal machines that were installed and to move the ore from place to place. The machines themselves were powered by water and so huge water wheels had to constructed to drive them. Machines to crush the ore cost a fortune and when all was said and done the crushed gossan sent away for assaying resulting in huge transportation costs. All this took time and as you know time is money.

Today, with the advantage of high gold prices that are continuing to rise due to the uncertainties of the present financial climate, a small scale commercial extraction facility could be a profitable concern. The machinery required for processing the gossam have become much smaller, are more efficient and not that expensive. These would be powered by diesel or petrol generators.

What would you need to establish a small extraction enterprise at the Bampfylde mine? You would need to establish a base camp from where you can operate the equipment and provide the necessary amenities for those employed by you. You will want to excavate the dumps on the western side of the Mole river because that is where the highest concentrations of gold is to be found. The problem is that these dumps are to be found at the top of a steep-sided hill.

There is vehicle (tractor, four-wheeled drive) access via a wide dirt track but the incline is very steep. In winter time and the wet season, I would doubt transportation would be practical. We also need water close by to be able to wash away the crushed ore during the extraction process. The river Mole would be the obvious source for the water. It makes sense therefore to locate the base camp at the same place where the original mine workings had been, near the ruined crusher house.

At the base camp you will require buildings that could be later removed or dismantled. I would suggest using a couple of mobile/holiday homes to be placed on the site, with basic electricity facilities, lighting, cooking (microwave), etc provided by electric generators. One could be used as an office, the other as a rest room with washing and cooking facilities. Second-hand 35 foot x 12 foot caravans are easily available at low cost. However, they should not be located where the Berdan machines had been housed. The ground was polluted by the mercury used by those machines. The cost would be circa £2000 each a total of £4000.

Electricity would provide the power to the buildings and machinery. In my estimate you would require about four diesel generators such as the

DE6800T 5.0kva 5 kva Electric Start Diesel Generator Super Silent Single Phase model that costs about £750 each. Two of these would provide power to the mobile/holiday homes and two to power the machinery used in processing the ore and for pumping water to these devices from the river. The cost would be £750 each a total: £3000.

You will need either to purchase or hire a small digger to transfer the gossan to a vehicle. The local farmers would probably be in a position to provide one at reasonable cost and as for the vehicle, this could be an open truck, a tractor pulling a cart, or another solution. It will only be used to transfer gossan from the west bank of the Mole river to the base camp on the east side. There is already vehicle access to the crusher house where the base camp would be situated. Norman Govier uses it to drop of sand for reconstruction. It would not be costly to improve this access as there is plenty of rock that could be used to strengthen the track. On the west side there is a large off-road area where the gossan could be loaded on the truck to be transported to the base camp.

These are just a few ideas that would make it possible to construct a small scale extraction facility, and all this can be done on a small scale with little environmental problems to worry about. It is just the case of removing and processing what is left of the dumps on the site and once done, nature will over time restore the land to what it once was.

What about the actual machines that will do the actual processing? The British Geological survey have suggested that the gold *"could be extracted using bulk concentrating techniques, e.g. using a Knelson Concentrator, without much pre-processing of the dump material except for screening to a suitable size."*

THE KNELSON BATCH CONCENTRATOR

The Knelson Batch Concentrator is widely considered to be the most technologically advanced gravity recovery batch system available today. It replaces all conventional gravity recovery equipment such as sluices, jigs, cones and spirals. The concentrator is commonly applied as a primary recovery device in precious metal milling circuits and alluvial treatment plants.

Knelson Concentrator - the low-cost and easy way to extract gold from ore

The batch model Knelson Concentrator operates on two basic principles; enhanced gravitational force and a patented material fluidization process.

1. Fluidization water is introduced into the concentrate cone through a series of fluidization holes.
2. Feed slurry is then introduced through the stationary feed tube and into the concentrate cone. When the slurry reaches the bottom of the cone it is forced outward and up the cone wall under the influence of centrifugal force.
3. The slurry fills each ring to capacity to create a concentrating bed. Compaction of the concentrating bed is prevented by Knelson's patented fluidization process.
4. The flow of water that is injected into the rings is controlled to achieve optimum bed fluidization. High specific gravity particles are captured and retained in the concentrating cone.
5. Upon conclusion of the concentrate cycle, concentrates are flushed from the cone into the concentrate launder through the patented multi-port hub.
6. Under normal operating conditions, this completely automated procedure is achieved in under 2 minutes in a secure environment.

At a cost ranging from £2000 to £6000 for a Knelson Batch Concentrator, together with diesel powered electrical generators and crushing machine a total investment would probably amount to no more that £20,000. That is a tiny price to pay for returns that could be as high as two million pounds excavating told at the the Bampfylde mine.

When the British Geological Society was on Bampfylde site in 1994 they set up a panning site on the River Mole. Here the samples they took were sieved at 2 mm and the washed oversized collected. The <2mm sample, approximately 1Kg in weight, was then panned to provide a heavy mineral concentrate of c 50g. Upon returning to their labs at Kenworth, the samples were dried and 16 sieved at 30 mesh (500 mm) and magnetically fractionated using a Franz Isodynamic Magnetic separator. Each magnetic fraction was subsequently super panned to concentrate the heavy minerals. Any gold grains were then hand picked and weighted.

A Franz Isodynamic Magnetic seperator could easily be installed at the Bampfylde mine to extract the gold. Alternately, since water is abundant at the location by virtue of the River Mole running through the area, one could even crush the ore almost to a powder and just pan for gold in the old traditional way. There are other gold extacting solutions too, such as wave tables, shaker tables or mineral concentration equipment such as those manufactured by Extracta-Tec. What all this boils down to is that with fairly low cost equipment which have a minimum footprint one could easily extract gold from the auriferous spoil lying on the surface of the Bampfylde mining area without too much difficulty. So what is the problem? Is it money?

I do not think that obtaining investment of say up to £100,000 would be difficult if potential investors were provided with the details described in this book. The problem is this. A great deal has changed since Victorian times when few people if any thought of the environmental impact of mining operations in the heart of the Devon countryside. Today, with ever increasing concerns about our environment, questions should now be raised whether a small to medium gold extracting facility at the Bampfylde site would cause too much damage to the environment? This issue was raised in 1974 as I shall now relate.

THE ALL IMPORTANT QUESTION

In 1974 planning permission was granted to a London-based firm called British Kynoch Metals Limited, who were acting on behalf of British Insulated Calender's Cables and Imperial Metal Industries to undertake a limited exploratory survey at the Bampfylde mine a view of mining copper.

Credit for the survey which made headline news in Devon goes to John Rottenbury. British Kynoch Metals, had sponsored him with a scholarship for him to further his work with Leeds University and through him, the company became interested in the Bampfylde mine.

COPPER MAY BE MINED IN DEVON AGAIN

1974

COPPER-mining could re-start within two or three years in a remote and scenic part of North Devon in an area near North Molton, Today all the country's veys over the past two years copper requirement are im- had made it worthwhile to ported at a total of £300m. and ment would th significant sav country's bala ments.

EXMOOR COPPER SEARCH MAY PROVE FRUITFUL

By MICHAEL SMITH, our City Editor

THE RESULTS of the copper search in Exmoor by British Kynoch Metals are now known. A cloak of secrecy surrounds the strength of the results of the pilot drilling programme, but they are understood to be highly encouraging.

1977

THERE NOW seems no chance of North Devon enjoying a copper boom, British Kynoch Metals, who have been test drilling in the North Molton and Molland areas for the past three years, announced this week they are ceasing any further exploration for copper in the the two areas.

The firm first took an interest in the area, which has been mined since Roman times, in 1972, when the world price of copper went sky-high. A geo-chemical survey at that time produced promising

But after a geophysical survey which has just been completed, the firm decided to abandon the project.

Had mining gone ahead it was expected to be underground.

Even so, for the past two years the Exmoor conservationist lobby has been preparing its big guns to fight any copper mining plans in the North Molton and Molland area.

1977

When British Kynoch Metals and B.I.C.C. had originally approached Sir Dennis Frederic Bankes Stucley, 5th Baronet (1907-1983) and his wife Lady Stucley the couple laid strong emphasis on the fact that if open-cast mining had been considered they would not have entertained it. They were absolutely opposed to any mining that would spoil the countryside. The company responded and stressed that they would maintain a responsible attitude to environmental considerations should economic deposits eventually be proved. They said that mining would be underground, not open cast, and the visual impact of equipment, plant and offices would be minimal. No local smelting was contemplated and that material would probably be trucked to the Midlands for smelting.

British Kynoch Metals was interested in the mine because at the time the world price of copper had gone sky high and all the UK's copper requirements was imported at a total annual cost of £300 million. To mine copper again profitably would make significant savings in the country's balance of payments. But it was not to be. On 18th February, 1977 it was announced in the press:

"There now seems no chance of North Devon enjoying a copper boom, British Kynoch Metals, who have been test drilling in the North Molton and Molland areas for the past three years, announced this week they are ceasing any further exploration of copper in the two areas."

The reason for withdrawing from the copper mining was not stated. But the issue here was that the mining had to be done underground and no open-cast surface working was permitted in case it damaged the environment. Today environmental issues are even more important. As far as I could see, if gold mining was to proceed at the mine, it would not be open-cast mining. What would happen would be the gradual removal of the gossan dumps from the site, which was already happening anyway. The farmers over the years have removed some for use as hard core for roads, stone wall fillers and other projects. Would Peter Stucley and his family see it that way? There was only one way to find out.

To open up the Bampfylde mine again is a truly a fantastic opportunity and it only requires someone with clarity of vision who recognises the potential of such an enterprise. As far as I am concerned one person comes to mind that may well fit the bill like a glove. His name is Peter Stucley. Not only does he own the mineral rights to the Bampfylde mine, a significant advantage, but also being only 40 years of age he is young and already runs the estate as a business enterprise, with farming and pheasant shooting, as revenue streams.

If I was in Peter's shoes I would jump at the chance., but the question is would he be interested in opening up the mine and to put his name to it. Of course he would require about £100,000 investment but even with this would he be able to do so in view of what happened in 1974. Would he or his family oppose such an enterprise?

At the end of June I received a call from Peter Stucley, following an email I had sent him about the book. He was very interested with what I had written and extremely keen to read it. I promised him that at the end of the following week I would send him a copy of the book in Acrobat format prior to it being published on Amazon for the Kindle. I anticipated that the book would be finished then. He said he would read it immediately.

During our brief conversation on the phone I could sense Peter's enthusiasm as I described some of the things I had discovered. Although he was aware of some information that I had uncovered, the results of the British Geological Survey came as a complete surprise. As for the sampling made by the Feltrim company which he knew had not got anywhere, he did not know that the results had been good and that the company could not pursue the matter further due to financial problems.

Referring to the environmental concerns and reservations of Sir Dennis and Lady Stucley in 1974 I asked Peter that if my book proved that gold was to be found in economical and commercial quantities, would he be interested in pursuing the matter further. Without divulging what he said to me in confidence, the answer was a positive yes! Could this book that I have written be the catalyst that will bring gold mining to Devon again? Watch this space...

EPILOGUE - 2013

THE PROPOSAL | THE SITE SURVEY | GOLD CONFIRMED | BOMBSHELL! | WHAT NEXT?

When I had finished my book which was published in July 2012, a few months later I contacted Peter Stucley who had kindly written the foreword for the book and asked him whether he was going to carry out any mining operations at the mines as a result of my researches. Unfortunately, he said he could not do so as he and his elder brother George were involved in other projects on the North Molton Estate which involved considerable investment and therefore any activity in the mining area was out of the question. However, he said that if there was somebody who wanted to lease the mining rights and therefore take the financial risk, then he would be interested to hear from them. I said that if I was ever contacted by anyone who would be interested in taking on the financial risk and investing in the mines I would let him know.

In November 2012, I received an email from Mr X who works for a number of companies involved in mining exploration and mineral extraction around the world. (I am unable to disclose his name or the mining companies for whom he represents due to a confidentiality agreement I have with him). He has been responsible for locating some of the largest deposits in Africa for his clients. He was being asked to review potential sites in the UK for a number of customers who had an interest in former operations and who have the ability to rework sites often abandoned due to economics of the time.

Mr X told me that the main reason gold mining in many parts of the world stopped at the first part of the twentieth century was simply because of economics. In those days gold was at a low of $20 an ounce or even less but labour costs and equipment had increased dramatically. It simply was uneconomic to extract gold under those circumstances. The situation had now changed. In the last year alone (2011) Mr X had reviewed former workings in Wales, Scotland, New Zealand, British Columbia, Ontario Canada and Nevada, USA all with the view of starting small operations again. This is has only become possible due to the rising gold prices and modern methods of metal recovery.

Mr X had been aware of the BGS report of 1994 and when he read my book he felt compelled to investigate further. He had done a great deal of large scale tailings recovery around the world and the description of the spoil at North Molton in my book suggested that a small scale extraction of gold at the old mines could be a viable option. This of course would be subject to a survey and the testing of samples, which he would undertake without cost to the owner of the land Peter Stucley. Furthermore, the recovery techniques employed were environmentally friendly and did not involve chemicals, water concentrators et al.

THE PROPOSAL

Mr X proposed that he would visit the site of the old mines at Heasley Mill for a few days and carry out a study of the present spoil dumps that were associated with the former workings. Tests would include using portable X-ray Fluorescence Spectrometer (XRF) analysis equipment that was also capable of rare earth detection.

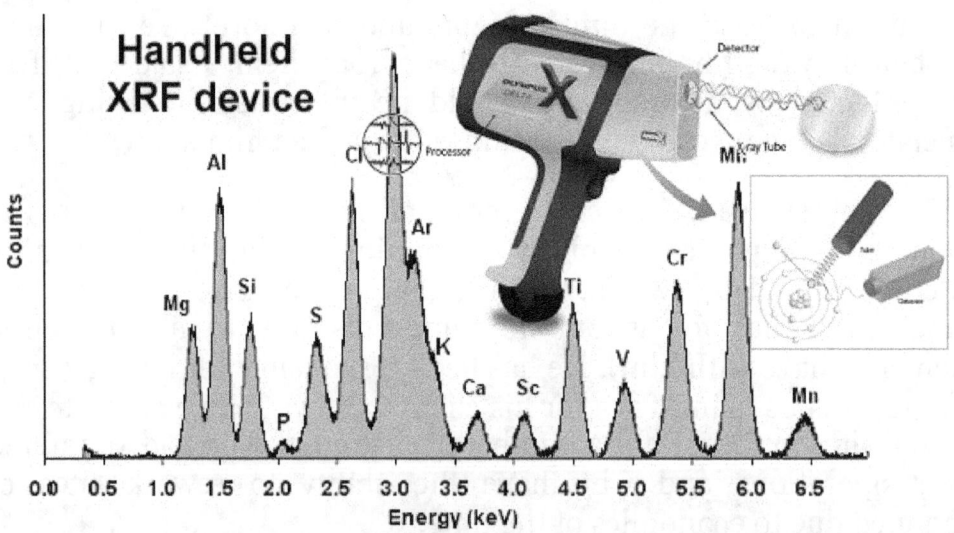

Hand-held XRF devices allowed Mr X to directly make an on-site geochemical analysis of rock samples and obtain results in a matter of minutes. These results are displayed on a computer notebook running special XRF software installed on a notebook computer which show a spectrographic analysis of what is tested. Of course there was more to this than just point and shoot with a XRF device. Mr X looked for pathfinders that were indicators for specific minerals. For example it was common to see high arsenic readings around good gold mineralisation, along with copper etc. He said that success in finding gold was all down to a combination of technology, science, geology and detective work.

Mr X's area of expertise involved micron and sub micron mineral recovery with special focus on precious metals and rare earth minerals. He had developed a technique for recovery from mine tailings and tailings ponds which is now seen as a preferred extraction method in alluvial and hard rock deposits with over a 90% recovery rate without any chemical requirements. The gold had to be in a native form and not an oxide although he did say that he had developed an environmentally friendly leaching process but his main work was still for native gold recovery down to nano particle levels. His lab was equipped with petrology, AAS - GF, Mass Spectrometer and XRF and he could offer a full range of analysis for rapid sampling.

Mr X said that test recovery of a number of samples (not panning) and detailed microscopic analysis of any metals present would take place at the mines. Any findings would be made available in report form to Peter Stucley and any gold recovered would be handed over to him after analysis. Based on these findings a proposal would then be made with an offer to recover the historic tailings, remediate any areas of contamination and make good the worked area. All mineral recovery would be completed on site and a full geological survey including drilling could be carried out on the possible opening of a new mine on the land. If the results of the survey was positive Mr X expected that bringing investment to the project would be straightforward with an agreed percentage royalty being granted to the land owner.

THE SITE SURVEY

Because Mr X travels all over the world in the pursuance of his work he was not often in the UK. Speaking on the phone in November 2012 he told me that he was anxious to visit the mines at North Molton as quickly as possible because he had only a few weeks left in the UK before heading to British Colombia to check on his latest tenure purchase in that Canadian State. All he needed to do was simply to access to the land because he had a complete field lab and he did not need anything else other than access. He said It would be necessary to carry out a complete a series of tests that included the use of XRF mining technology on the various tailings around the former workings to see if the results were worthy of taking further, as my book had suggested.

I passed on the details to Peter Stucley who thanked me for introducing Mr X to him. I then waited with bated breath for news on how the survey panned out as Peter had said to me that he had agreed to allow Mr X to visit the site and carry out a survey. I heard nothing more for a couple of weeks then I learned that the survey had been postponed to early next year (2013). The reason for this was because of the atrocious weather that the country had been experiencing. It just rained, rained and rained. According to the Met Office, 2012 was the wettest year in England since records were kept in 1910 and Devon had not escaped the 'monsoon'. Devon saw its second wettest year with 1,607.7mm (63.3in) of rain. Carrying out a survey under such conditions was simply out of the question.

I heard nothing more but ever curious I contacted Mr X at the end of January enquiring if any date had been set for carrying out the survey. He replied that he was currently in Africa and that no plans had been discussed as to when he would be able to do the survey. I left it at that and did not think any more about it.

Six months was to pass before I once again thought about the Bampfylde mines at Heasley Mill. The summer of 1913 had been glorious and it seemed that it would have been a great opportunity for Mr X to carry out his survey, but I had not heard anything more about it. What really, set things in motion was that I had been approached by a gentleman who was interesting in filming at the old Bampfylde Mine as part of a Channel 4 programme that his company was making. Needless to say I passed on his details by email to Peter Stucley and at the same time I asked if there had been any progress with the survey.

I received a reply from Peter thanking me for passing on the details of the television company but for some unknown reason he did not answer the question which I had asked him. I thought this was very odd because he had always responded to any questions that I have raised in the past concerning the mine. Perhaps, I thought, Mr X had decided not to do a survey and therefore there was nothing more to be said.

BOMBSHELL!

My curiosity aroused I decided to email Mr X and ask him if he was still going to the survey. When he replied, what he said could have knocked me down with a feather. He had already carried out a detailed survey and sampling of the site over a period of four days and had reported his findings to Peter. The results, he said, were conclusive. Economic gold recovery was possible from the tailings and spoil heaps of the former copper mine.

I was astonished! I wrote to Peter about what I had learned but he did not respond to my email. So I contacted Norman Govier who had helped me so much in the writing of my book and for showing me around the mines. I wanted to let him know what I had discovered. I spoke to Norman on the phone and was astounded to learn that he had been present when Mr X carried out his surveys. The reader can imagine how I felt being left out of the equation. However, if Peter wanted to keep the survey secret from everyone, including myself, then that was his prerogative. Peter must have had his reasons and there is no point dwelling on it any further. The deed had been done and one cannot turn back the clock.

GOLD CONFIRMED

Now that the "cat was out of the bag" Norman opened up and described how the surveys were carried out during the four days of investigation. He described the equipment Mr X used and how it confirmed that gold had been found in the spoil in commercial quantities, just as I said it would in my book. Norman had always doubted that the rumours of gold being found at the Bampfylde mines, but now there was no doubt. Later, Norman even dropped round a sample of the gold taken at the survey to show me. The microscopic gold was at the bottom of a tube, but it was gold nonetheless. My investigations had born fruit. I had proven through careful investigative journalism and a lot of luck that commercial quantities of gold is to be found at the Bampfylde mines. What I had written had been vindicated.

What next? Having confirmed that there was gold in sufficient quantities to make it worthwhile to extract, Mr X had suggested a commercial model to recover it and this was all very agreeable to Peter. The papers were about to be drawn up and licensing signed when there was an unexpected twist. It transpired that Peter did not own the mineral rights after all. His elder brother did and George did not want any part in mining the land for gold. What can I say?

I suspect that the pheasant enterprise in the mining area that both Peter and George share and which brings in a good revenue to the estate was far too important to risk any mining disruption in the area, even though Mr X had said his activities would be non-intrusive. For the extraction of gold to take place it would have meant that the pheasant shooting would have had to be stopped as it would have been too dangerous for Mr X and his associates to work with bullets flying around.

Not surprising, Mr X was not happy with this revelation, having lost four days of time and money on the site and a great deal of lab analysis in expectation that if he found gold that an agreement could be signed and the gold extracted. But as Mr X said to me in an email, he went into this project like any other speculative visit and he has now moved on. For me I was very disappointed. I had been an outsider who had moved to Devon in 2009 and yet, I had uncovered about £2 million worth of gold in the tailings of the mine, when nobody had thought this was possible for over a hundred years or more. Needless to say I was gratified at my discovery but disappointed that an unexpected chain of events has now prevented the extraction of the gold.

WHAT NEXT?

What now? One can understand why today the Stucley family are keeping news of the gold on their land a secret. The last thing they will want is for trespassers trampling over their property in search of gold. In any event, such prospecting would be in vain. The gold exists in almost microscopic particles within the spoil littering the land, which means in order to extract it, the rocks have either to be crushed or some other modern extraction technology would have to be involved. One thing is for certain, there are no gold nuggets to be found. Furthermore, the area of the mines can be a dangerous place to visit.

For example, from 1st October to 1st February, the land is a war zone. This is the game season and the Stucley family operates a commercial game-bird shoot on the land that is leased to a syndicate. And as for the mine shafts themselves. Norman tells me that they have only been partially filled in. It would have been impossible to fill in the shafts completely, they were too deep, so a false ceiling of cross beams was place some way down the shafts and filled with rubble from the top. This means that beneath them, the mine shafts are still open. Should the beams ever fail, there could be a disaster in the making.

Then of course, there is the area where the Brendan machines had operated. It is a small area, but it is contaminated with mercury. Yes, the mining area could be a deadly trap for the unwary. You would not want to picnic at this location.

As for the gold at the mines of North Molton it is still there, but the question is for how long? Norman says that the Stucley estate is removing the spoil to build walls and roads. How ironic it would be if a rambler was to one day traverse the many footpaths in the vicinity of Heasley Mill, and sees the glitter of gold shining within some farm walls. I wonder if he would realise that what he was looking at was some gold from the mines. Who knows? One day it could be said that the roads of North Molton are paved with gold - and it would be true.

I end this book with a warning that you ignore at your peril. The Bampfylde mine area is private property, extremely dangerous, and as such access is restricted to authorised personnel only.

BIBLIOGRAPHY

- **Anon** *History and Description of the New Bampfylde Copper Mine, Exeter* (prepared for the British Association meeting at Exeter 1869).
- **Claughton, Peter**, *Gold at North Molton - and the surviving evidence to be found on the Bampfylde Mine site, Devon.* (Dept. of Economic and Social History, Exeter University, 26 September 1998)
- **Cameron D G, BSc, BA and Bland D J, BA.** , *An Appraisal of the Gold Potential of Mine Dumps in the North Molton Area, North Devon* (British Geological Survey © NERC. All rights reserved. CP12/053, 1994)
- **Cornwall Record Office**
- Devon Record Office (formerly Exeter City Record Office)
- Durrance and Laming, The Geology of Devon
- Guildhall Library, Stock Exchange Loan and Company Prospectuses, The Britannia Gold and Copper Mining Company, May 1852; The Poltimore Copper and Gold Mining Company, undated.
- Kahlmeter's Journal, Kungliga Biblioteket, Stockholm - M249 Vol. 3 H. Kahlmeter, Dagbok ofver en ...... 1718-26 foretagen resa .
- Mining Journal (MJ)
- North Devon Journal (NDJ)
- Pattison, S R. A day in the North Devon Mineral District, Trans. Roy. Geol. Soc. Cornwall, vol. VII (1853), pp. 223 - 227.
- Public Record Office (PRO)
- Rottenbury, J. Geology, Mineralogy and Mining History of the Metalliferous Mining Area of Exmoor, unpublished Ph.D. thesis, Leeds, 1974
- Slader, J M. Days of Renown, 1965.

ABOUT THE AUTHOR

I was born at Pittsburgh, Pennsylvania in the USA when my English mother was visiting the country. Having returned home to England I was brought up and adopted. I have never been to the USA since I was born. As a result, I am a British non-fiction author who enjoys investigating mysteries and writing about them. This present work came about by accident when I came upon a small sentence in a local village book, that suggested that there was gold at the mines of North Molton. The mystery was that nobody seemed to know much if anything about it, not even my local museum in Barnstaple. It was a mystery I wanted to solve and I did.

The greatest mystery of all is the question of life, the universe and everything, and it has been a lifelong passion for me. Throughout my life I have studied Religion, Palaeontology and the Life Sciences and as a result I can be classed as an Old Age Creationist. I am a keen student of the Bible and Archaeology, and I have studied many ancient writings, often in their original languages. I am also a keen researcher on Biology, Genetics, Theology, Mythology, Geology and the Supernatural, and often you will find in my books that they incorporate many of these sciences in their content.

OTHER BOOKS THAT I HAVE WRITTEN

NEPHILIM SKELETONS FOUND

This book has been a big hit, especially in the USA, and since it was published in December 2013 for the Kindle, it has been in the top ten in the category of Paleontology, often in the first five positions and it has been #1 on a number of occasions.

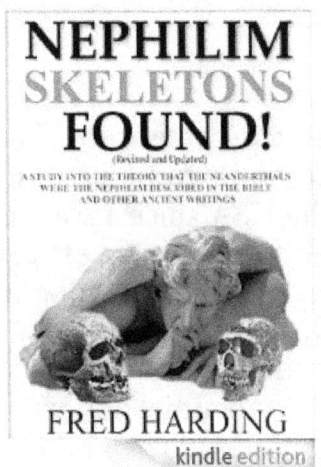

Nephilim Skeletons Found [Kindle Edition]
Fred Harding (Author)
(41 customer reviews)

Super research - surprising conclusion.
By valve timer
Excellent and complete research using the words and publications of evolution proponents to prove his points.

Blew my mind
By DeAndre Tinker
Incredible book. I has been toying with some of these ideas in my head and he articulated them and took them much further. Essential reading for anyone interested in history.

Well thought out
By Tom Rudolph
Research and solid thinking really opened my eyes to the value of the Book of Enoch. Everyone should read it.

A remarkable book
By Claire
This is a remarkable book to say the least. Harding's scholarship is outstanding. He has done a marvellous job in pulling together the words of leading proponents of evolution as found in their own works such as press releases, newspaper accounts, science journals, drawings and speeches and through them, without referring to any outside anti-evolutionist commentaries, disproved their own theory. I have never seen anything like this before. I cannot imagine how anybody, upon reading this book, could possibly believe that man had evolved.

This book is a comprehensive study that shows that the skeletons attributed to Neanderthals were in fact those of the Nephilim described in the Bible, the hybrid offspring of angels and women.

AMAZON UK
Kindle
http://www.amazon.co.uk/Nephilim-Skeletons-Found-Answer-ebook/dp/B00CKU8I5M
Paperback
http://www.amazon.co.uk/Nephlim-Skeletons-Found-Fred-Harding/dp/1500702323
AMAZON USA
Kindle
http://www.amazon.com/Nephilim-Skeletons-Found-Answer-ebook/dp/B00CKU8I5M"
Paperback
http://www.amazon.com/Nephlim-Skeletons-Found-Fred-Harding/dp/1500702323"

THE TIMES OF THE GENTILES ARE FULFILLED

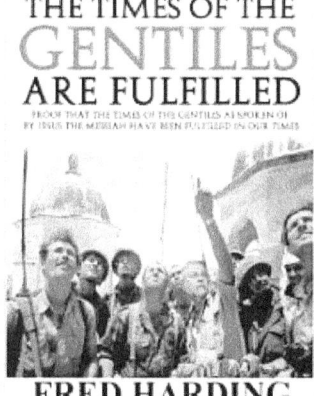

This is the most historically accurate and comprehensive study on the subject ever produced incorporating some remarkable discoveries. It is written for both Jews and Christians alike, together with an understanding of how Islam figures in the scheme of things. Once your read this, which is supported by verifiable evidence, your view of the world may never be the same again.

TO BE PUBLISHED FEBRUARY 2015

EVOLUTION'S COUPE DE GRACE

This is a study that shows how man's almost hairless condition and his remarkable penis proves that Man could not have descended from an ape-like common ancestor. These human attributes are the most difficult subjects that Darwin and those that adopted his theory of Natural Selection have ever tried to answer. In fact it is the one problem that Darwin's theory, in whatever evolutionary guise it appears today, has been unable to explain.

EVOLUTION'S Coup de Grâce [Kindle Edition]
Fred Harding (Author)
☆☆☆☆☆ (4 customer reviews)

- Length: 199 pages (estimated)

★★★★★ Excellent book, well worth a read
By Jules

I have read a number of Fred Harding's books and have to say that this is one of his best works. He tackles a somewhat taboo subject of the evolution of the Primate penis in a scientific way, looking at all the evidence, which creates a most compelling argument for Intelligent Design.

★★★★★ Darwin is hairless and hydraulic
By Brian2520

Well Harding has most eloquently demonstrated that the Darwinian Emperor and all his entourage are not wearing any clothes and would you believe it, they aren't covered in insulating fur and their gentalia even operate on hydraulics! Very un-ape like indeed!

★★★★★ A Great, Convincing Read
By R. Wolfe

This is yet another densely informative work from Fred Harding. Evolution's Coup de Grâce is a methodical yet interesting read that takes the reader through a clear, well-presented argument for how a certain piece of Man's anatomy and that of our supposedly closest evolutionary relatives are different enough to poke a great big hole in evolutionary theory. In short, this is a fantastic read that fires a scientific, well-informed, and brutally logical shot right through evolutionary theory.

Charles Darwin declared in his famous book *On the Origin of Species*:

"If it could be demonstrated that any complex organ existed, which could not possibly have been formed by numerous, successive, slight modifications, my theory would absolutely break down. But I can find out no such case."

Likewise, Dawkins explains in his book *The Blind Watchmaker* that any incredulity about any incredibly complex transformation can be explained "*only if we stress that there was an extremely large number of steps along the way, and if each step is very tiny.*"

So according to Dawkins, the key to explaining how something as complex as the human penis being transformed the way it has is to recognise how natural selection enabled the innumerable tiny random mutations gradually weeded by the environment was able to produce astonishingly complicated self-replicating systems.

Well Mr Darwin and Mr Dawkins, there is a complex organ that could not be possibly have been formed by numerous, successive, slight modifications and that is the human penis. As will be shown in this book, your explanation cannot explain how the common ancestor's penis changed the way it did, so that humans have a completely different type from those of our hypothesised living ape cousins (chimpanzees and bonobos) that is supposed to have branched from the same common ancestor.

As the study unfolds, you will learn how evolutionists have not been able to explain how Man became hairless and how the anatomy of the human penis completely disproves the belief that Man had evolved from common ancestors that links him to the apes. The time has now come to bring the penis out into the open, so to speak, and drag the theory of evolution to where it belongs. Oblivion!

Hence, in Darwin's own words, "my theory would absolutely break down" - and it does big time!

AMAZON UK
Kindle
http://www.amazon.co.uk/EVOLUTIONS-Coup-Gr%C3%A2ce-Fred-Harding-ebook/dp/B00GOFPB2M <
AMAZON USA
Kindle
http://www.amazon.com/EVOLUTIONS-Coup-Gr%C3%A2ce-Fred-Harding-ebook/dp/B00GOFPB2M

BREAST CANCER: CAUSE - PREVENTION - CURE

When a friend died of breast cancer, this was another mystery I felt that needed to be investigated. There was no breast cancer in her family so why did Gill die of this terrible disease. So I carried out extensive research on the subject and discovered not only what causes breast cancer and to prevent it but I discovered a cure that, is as I write this, is undergoing human trials by an independent pharmaceutical company.

Breast Cancer: Cause - Prevention - Cure
[Kindle Edition]
Fred Harding (Author)
★★★★★ (2 customer reviews)

★★★★★ **A Must read for everyone !!**
By **Simone Simon**
Absolutely amazing so detailed and puts everything about breast cancer in to perspective. Everyone should read this book !!!!

★★★★★ **Women, Breast Health Must Read!,**
By **Judy Krings, Ph.D., CMC, PCC**
Having lost a friend to breast cancer, Harding's book is a legacy to her. An intense, rich labor of love. I haven't read anything as fascinating in a long time. I didn't plan to read it in one sitting, but it grabbed me and took me to places as a woman I had never been. Harding wove women's history, culture, science and artistry into one life-saving book. You think your bra is your best friend? Guess again! This book skillfully opens Pandora's box to dirty little secrets that are killing women. Watch out certain pharmaceutical companies. Read this book and save your life.

The synthetic version of a natural herb that is being developed cures all cancers too. Read more about the research by following this link.
http://www.artbiomedical.com/

I discovered too that the cancer charities have no desire to find a cure and have in fact fooled the public in obtaining millions of dollars of research funds, where in fact very little goes on research to the causes, only treatment. If a cure was found their multi-million enterprises would be finished and they know it. I had opened up a can of worms. Since the publication of the book, what I have written has more than been vindicated.

AMAZON UK
Kindle
http://www.amazon.co.uk/Breast-Cancer-Cause-Prevention-Cure-ebook/dp/B005HM9BHY
Paperback
http://www.amazon.co.uk/Breast-Cancer-Cause-Prevention-Cure/dp/0955422108
AMAZON USA
Kindle
http://www.amazon.com/Breast-Cancer-Cause-Prevention-Cure-ebook/dp/B005HM9BHY
Paperback
http://www.amazon.com/Breast-Cancer-Cause-Prevention-Cure/dp/0955422108

CAREER HISTORY

I have been involved in IT since 1983 and between 1990 - 1995 I was Technical Director of a PC manufacturing company. My particular interest though has been in software development and from 1997 - 2009 I was Software Development Manager for a major health and safety organisation. In 2003 I became the finalist for the Oscar like BCS awards for software excellence for the IT industry with a COSHH Mananagment System that I wrote for the company. As a consequence of my computer programming experience, I tend to apply logic and reason to any subject that I study and so my books are not based upon any preconceived ideas.

I think what the famous long time atheist Antony Flew (1923-2010) said before he died having discovered God says it all, "I have followed the argument where it has led me. And it has led to accept the existence of a self-existent, immutable, immaterial, omnipotent, and omniscient Being".

In order to write books for the Kindle I have written a software product called Kindle Writer and it is now being sold all over the world, helping authors write books for the Kindle. You can find out more at www.kindlewriter.co.uk.